Filter Design Solutions for RF systems

Filter Design Solutions for RF systems

Editors

Leonardo Pantoli
Vincenzo Stornelli

MDPI • Basel • Beijing • Wuhan • Barcelona • Belgrade • Manchester • Tokyo • Cluj • Tianjin

Editors
Leonardo Pantoli
Università degli Studi dell'Aquila
Italy

Vincenzo Stornelli
Università degli Studi dell'Aquila
Italy

Editorial Office
MDPI
St. Alban-Anlage 66
4052 Basel, Switzerland

This is a reprint of articles from the Special Issue published online in the open access journal *Electronics* (ISSN 2079-9292) (available at: https://www.mdpi.com/journal/electronics/special_issues/filter_design).

For citation purposes, cite each article independently as indicated on the article page online and as indicated below:

LastName, A.A.; LastName, B.B.; LastName, C.C. Article Title. *Journal Name* **Year**, *Article Number*, Page Range.

ISBN 978-3-03943-547-0 (Hbk)
ISBN 978-3-03943-548-7 (PDF)

Contents

About the Editors

Leonardo Pantoli is a Tenure-Track Researcher and Professor with the University of L'Aquila (IT) and Vice-President and founder member of the spin-off SENSing s.r.l. He received a Degree (cum laude and mention) in Electronic Engineering and a Ph.D. in Electrical and Information Engineering from the University of L'Aquila, L'Aquila, Italy, in 2006 and 2010, respectively. In 2007 and 2008, he spent several months with the "Dpto. Ingenieria de Comunicationes—ETS de Ingenieros Industriales y de Telecomunicación" of the University of Cantabria, Spain, and the "C2S2 Department" of the XLIM Research Institute, Brive La Gaillarde, France. From 2013 to 2017 he was Research Assistant with the University of L'Aquila. From 2017 to 2019 he was Researcher (Law n. 240, 30 December 2010, Art. 24, letter a) with the same University, and in 2017 he obtained a National Scientific Qualification (Law n. 240, 30 December 2010, Art. 16, paragraph 1) to serve as Associate Professor in Italian Universities in the sector 09/E3—Electronics. Since 2019 he has been Tenure-Track Researcher (Law n. 240, 30 December 2010, Art. 24, letter b) with the same University of L'Aquila. His research activities include the development of methods and algorithms for the design of RF, microwave and millimeter wave nonlinear circuits, the stability analysis of circuits in both linear and large-signal regimes, active filter design, and MMIC design for aerospace, wireless communication and imaging applications. He has good experience with GaAs, Si and SiGe technologies.

Vincenzo Stornelli was born in Avezzano, Italy. He received the "Laurea" degree (cum laude) in electronic engineering in 2004. In October 2004, he joined the Department of Electronic Engineering, University of L'Aquila, L'Aquila, Italy, where he is involved as an Associate Professor. His research interests include several topics in computational electromagnetics, including microwave antenna analysis for outdoor ultrawideband applications. He serves as a reviewer for several international journals and as an Editor of the Journal of Circuits, Computers and Systems.

Preface to "Filter Design Solutions for RF systems"

Nowadays, technology developments and system integration capabilities are leading to the definition of innovative architectures and modules that require the re-design of many electronic components. Among them, one of the more critical at the system level is filters. In practice, these are difficult to integrate, and in many cases they perform poorly due to the technological limitations of passive elements. Tunability and calibration are often difficult to achieve as well. Different solutions are currently available on the market, and these have already been presented in the literature, depending on the operational frequencies and applications. For instance, modern communication systems have strict performance requirements from electronic devices and components, and these requirements lead to the choice of the technology solutions and component topology to adopt.

Filters can be designed at the electrical, mechanical or electromechanical level; they can be conceived as discrete or distributed components, and among the electrical ones, they can be realized with passive or active circuits. Furthermore, tunability is often a desired characteristic, especially in modern re-configurable systems, but it is usually difficult to achieve. It can be obtained in mechanical filters with screws driven by micro-motors, and in electrical ones by means of voltage-controlled variable components. In any case, it is not trivial to obtain a good tunability and to preserve the same shape factor in the full tuning bandwidth. A further criticism relates to the calibration of these filters, mainly at the industrial level, since the complete characterization of these components is usually obtained with a trial-and-error approach, but this is dependent on the expertise of the technicians. In this regard, automatic test systems optimized for the calibration of the device under test might be very fruitful.

This Special Issue, named "Filter Design Solutions for RF systems" (https://www.mdpi.com/journal/electronics/special_issues/filter_design) will focus on the state-of-the-art results in the definition and design of filters for low- and high-frequency applications and systems. Aspects related to both the theoretical and experimental research in filter design, CAD modeling and novel technologies and applications, as well as filter fabrication, characterization and testing, are covered.

Leonardo Pantoli , Vincenzo Stornelli
Editors

 *electronics*

Article

Low-Current Design of GaAs Active Inductor for Active Filters Applications

Leonardo Pantoli [1], Vincenzo Stornelli [1,*], Giorgio Leuzzi [1], Hongjun Li [2] and Zhifu Hu [2]

[1]   Department of Industrial and Information Engineering and Economics, University of L'Aquila, 67100 L'Aquila AQ, Italy; leonardo.pantoli@univaq.it (L.P.); giorgio.leuzzi@univaq.it (G.L.)

[2]   Hebei Semiconductor Research Institute, Shijiazhuang 050000, China; mmc-lhj@hbmmc.com (H.L.)

*   Correspondence: vincenzo.stornelli@univaq.it

Received: 30 May 2020; Accepted: 23 July 2020; Published: 31 July 2020

Abstract: Active inductors are suitable for MMIC integration, especially for filters applications, and the definition of strategies for an efficient design of these circuits is becoming mandatory. In this work we present design considerations for the reduction of DC current in the case of an active filter design based on the use of active inductors and for high-power handling. As an example of applications, the approach is demonstrated on a two-cell, integrated active filter realized with p-HEMT technology. The filter design is based on high-Q active inductors, whose equivalent inductance and resistance can be tuned by means of varactors. The prototype was realized and tested. It operates between 1800 and 2100 MHz with a 3 dB bandwidth of 30 MHz and a rejection ratio of 30 dB at 30 MHz from the center frequency. This solution allows to obtain a P1 dB compression point of about −8 dBm and a dynamic range of 75 dB considering a bias current of 15 mA per stage.

Keywords: active filters; active inductor; MMIC; tunable filters

1. Introduction

On-chip passive filters are affected by the limited Q-factor of inductors and capacitors, due to ohmic and substrate losses, even on low-loss substrates as Gallium Arsenide. Tunable passive filters are also affected by the limited Q of varactors, normally used for tunability. Moreover, the bandpass of the filter is affected by the combination of constant passive inductance and variable capacitance in the tuning range. Active filters can be realized with several approaches [1–3]; many of them are usually based on active inductors (AIs), that can achieve very low or even negative equivalent resistance, and therefore a high filter Q. The AI can be tuned both in terms of equivalent inductance and of equivalent resistance, yielding constant bandpass with limited losses or positive gain.

In general, active filters usually have limited power handling capabilities also due to the nonlinearities of the active elements, and they are prone to instability due to the negative resistance required for the compensation of the losses of the passive elements in the circuit.

In order to increase the dynamic range, usually a high bias current is required for the active devices. This can lead to higher power consumption that is often unacceptable for integrated circuits and also not allowed at system level. A possible means to reduce the bias current is provided by Class-AB bias, which has already been demonstrated [4]. However, this approach is not always possible or, at least, it experiences some drawbacks depending on the characteristics of the technology process (e.g., the availability of complementary transistors, highly linear active devices).

In this paper, we present a design approach of integrated active inductor and its applications in filters realized in GaAs technology that allows the minimization of bias current, still maintaining Class-A operations that usually ensure high-power handling capability. Concerning the filter AI base design, the proposed approach is based on AIs coupled with shunt capacitors in order to realize an equivalent high-order filter with good performances in terms of shape factor and dynamic range [5–8].

The topology of the AI is such that only a fraction of the bias current is effectively drawn by the active device and it is useful to provide the required negative resistance. In addition, each cell makes use of a single transistor, reducing power consumption and minimizing possible instability concerns.

An example of application is proposed in MMIC technology, potentially tunable in the frequency range between 1.8 to 2.1 GHz, with a tuning bandwidth of about 15%. The −3 dB bandwidth is almost constant and equal to 30 MHz, while the out-of-band rejection is significant thanks to the high shape factor equal to 2.5 for a 30 dB/3 dB bandwidth ratio. The chip has been designed with a standard process provided by HSRI Foundry that implements 0.13 μm GaAs pHEMT devices. Both simulations and on-chip measurement results are presented with a good agreement between them.

Currently; however, no varactor diodes are available in this technology. Therefore, a fixed-capacitor version was implemented, with different values of the capacitors. In fact, our aim is to provide a feasibility proof demonstrating the tunability capability of the filter by replacing the unavailable varactors with fixed capacitors. A version with the varactors will be implemented and fabricated as soon as the technology is available. However, good performances with varactors have already been demonstrated in a hybrid implementation [5], that indicate the possibility of a successful implementation with varactors also in monolithic technology.

Noise figure has not been considered in this design since noise performance requirements were not so strict in the proposed application (as later discussed); therefore, it is quite high. However, noise reduction techniques have been developed and patented [7] that yield a relatively low noise figure. They will be dealt with in a future publication.

In the following, the topology and principle of operation of the active inductor is briefly summarized in Section 2. Section 2 illustrates also the proposed low-current design approach on an active filter, while in Section 3 the MMIC design and measurement results are shown. Finally, the conclusions are drawn (Section 4).

2. The Active Inductor Architecture

Active filters can be realized by one or more cells, each including an AI-based shunt resonator (Figure 1). Both the shunt capacitor and the AI are tuned in order to maintain both insertion loss and bandpass almost constant across the tuning range. Thus, it is straightforward to notice that the AI is the centerpiece of the filter, as it is also able to control the losses of the cell and; therefore, the overall quality factor of the filter.

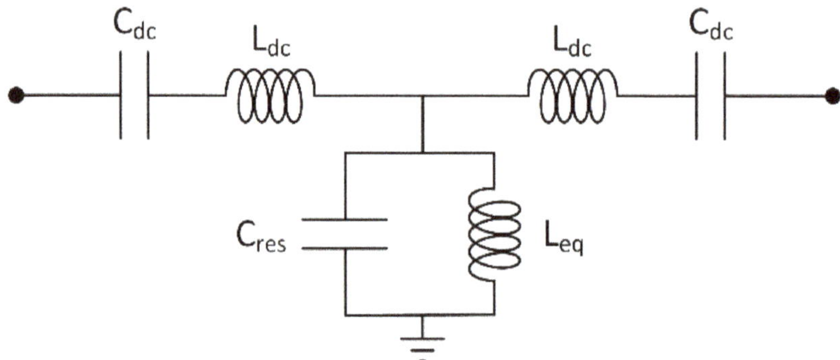

Figure 1. Topology of a single cell for bandpass active filters.

The traditional structure of the AI is shown in Figure 2a. The input voltage is sampled by a non-inverting transconductance amplifier, that drives a current into the capacitor. The voltage generated in the capacitor is sampled by the inverting transconductance amplifier, that draws an inductive current from the input of the active inductance. The relations between voltages and currents in the active inductor can be better understood in the complex phasor plane (Figure 3). Phases are referred to that of the input voltage V_{in}. The capacitor voltage V_C has a 90 degrees delay (capacitive delay) with respect to the input voltage and the inverting transconductance introduces a further 180° phase shift, so generating an inductive current I_{ind} with respect to the input voltage V_{in}. An overall excess phase of the output current with respect to the purely inductive 90° phase shift gives an equivalent negative resistance in addition to the equivalent inductance, while an insufficient phase shift gives a positive equivalent resistance. The amplitude of the inductive current with respect to the input voltage determines the value of the equivalent inductance. This amplitude can be changed, for instance, by tuning the value of the transconductance(s).

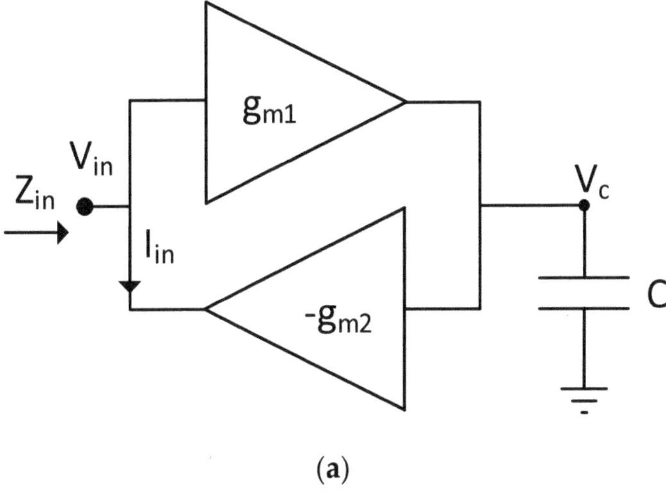

(**a**)

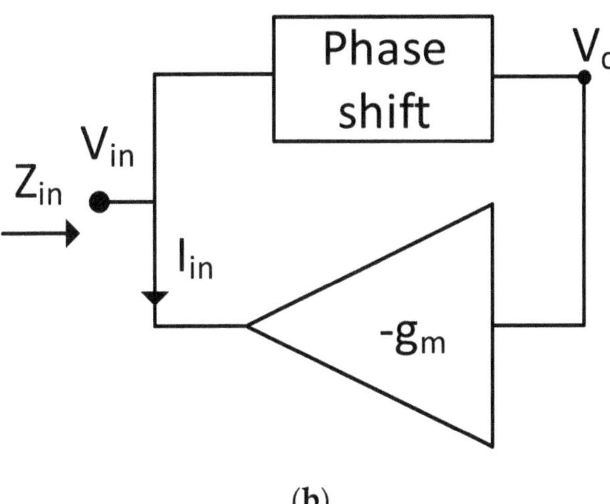

(**b**)

Figure 2. Active inductor architecture: (**a**) Traditional topology, (**b**) improved topology.

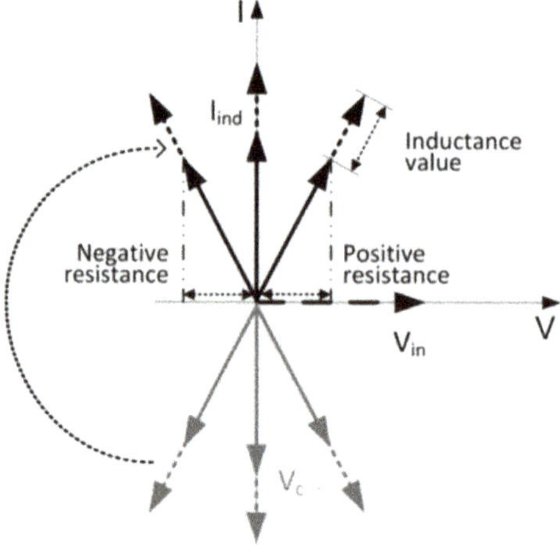

Figure 3. Phasors of voltages and currents in the AI of Figure 2 in the complex phasor plane. Vin is the input voltage, V_C the current across the capacitor (Figure 2a) or at the output of the phase shifting network (Figure 2b), and I_{ind} the current drawn from the inverting transconductance amplifier.3. Low-current design of GaAs active filters based on the proposed AI.

An improved topology of the AI is shown in Figure 2b [5]. The input voltage is sampled at the input of a phase-shifting passive network, then transferred to the input of the inverting transconductance amplifier, that in turn draws the current from the input port. The relations between voltages and currents are approximately the same as in Figure 3; however, some phase shift is also introduced by the inverting transconductance amplifier, given the high operation frequencies. In this improved topology, the amplifier is a fixed-bias, class-A linear amplifier, stable and with fixed gain. The tuning of the phase and amplitude is; therefore, performed by the phase-shifting passive network that includes varactors. This approach allows one to obtain stable and easily tuned AI, and consequently filters with a relatively high gain compression point.

For relatively high-power handling, the current in an AI, conceived with a class-A polarization, may become relatively high. With class-A operation, this requires a high bias current in the active device, where the whole output current flows. A possible means of reducing the DC current is the use of a class-B or class-AB amplifier in the AI [4]. This is certainly a feasible approach that significantly reduces the DC power requirements of the filter, even if it may cause some concerns on intermodulation. However, this approach is not always possible, depending on the characteristics of the active device. For instance, a very sharp pinch-off of the transistor makes the class-B or AB impractical; therefore, a different approach is here proposed and successfully applied to the design of an active filter.

In order to have a high-Q filter, the current at the input of the active inductor must be purely inductive, or nearly so, that is, with a 90 degree phase relation with respect to the voltage across it. The current through a passive inductor has less than 90 degrees with respect to the applied voltage, due to losses. This current can be seen as the vectorial sum of a purely inductive current, and of a purely resistive current in phase with the applied voltage, much smaller in amplitude, due to the losses in the inductor. The compensation of the losses can be obtained by summing another current, opposite in phase with respect to the resistive current and; therefore, equivalent to the current of a negative resistance. This negative resistance current will be therefore much smaller than the total current in the inductor.

A possible topology that implements this approach is shown in Figure 4. The current from the input of the AI flows through a passive inductor (I_L in Figure 4), because the capacitive current that enters the gate of the active device is relatively small and can be neglected. Then, the main part of this current ($I_{passive}$ in Figure 4) flows through a relatively large capacitor C_g, that has a smaller impedance compared to the passive inductor. Therefore, this current is mainly inductive with respect to the input voltage, but also has a resistive component, due to the losses in the capacitor and in the inductor. From the plot in Figure 5 its resistive nature is apparent from its phase relation to the input voltage.

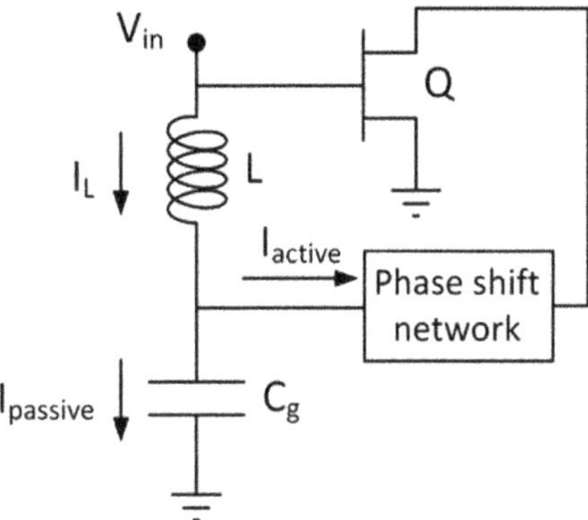

Figure 4. Improved topology of the AI for class-A low bias current.

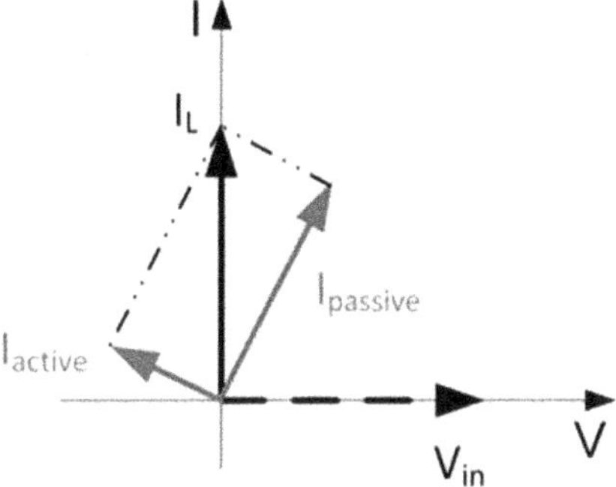

Figure 5. Phasors of voltages and currents in the AI of Figure 3 in the complex plane. V_{in} is the input voltage, I_L the current through the inductor L. $I_{passive}$ is the current through the capacitor C_g and I_{active} the current drawn from the transistor through the phase-shifting network. Their vectorial sum gives the current through the inductance I_L.

A smaller part of the current through the passive inductor flows through the phase-shifting network (I_{active} in Figure 4). This current can have a negative resistance component with respect to the input voltage, because it ultimately comes from the active device. The size of the active device, and the delay introduced by the phase-shifting network must be designed in such a way, that the negative-resistance component of the I_{active} current compensates the resistive component of the $I_{passive}$ current through the capacitor. In this way its amplitude is kept to a low value compared to the total current flowing into the AI. Therefore, the current through the active device is minimized, compared to the standard design, where all the current flowing into the AI comes from the active device. As a consequence, bias current and overall power consumption in minimized.

The value of the equivalent inductance is easily tuned by varying the value of the capacitance C_g, that can be implemented with a varactor; given the relatively high value of the capacitor, the series L - C_g is inductive, because the operating frequency is higher than the resonant frequency of the LC series. By changing the value of the capacitor, also the equivalent inductance of the LC series is varied. Another varactor can be used also to implement the resonating capacitance C_{res} of the single cell (Figure 1). The simultaneous tuning of the resonating capacitance and of the equivalent inductance yields a constant bandpass across the tuning frequency range, while the low losses or possibly small negative equivalent resistance of the AI allow to enhance the Q-factor of the filter.

The proposed design approach has some similarities with the negative impedance converter (NIC) approach [9,10]. However, the negative resistance is not designed as a one-port, additive network, but is obtained by suitable phasing of the active inductor loop. Moreover, the very simple AI topology here addressed, based on a single transistor per cell, greatly reduces the current requirements, and makes stability enforcement quite straightforward, ensuring at the same time also high-power handling together with low-power consumption.

When properly implemented, the proposed approach does not cause any reduction in the tuning range or increase in losses with respect to the traditional approach. The main result is the reduction of the current required from the active device, with consequent increase of power handling with the same active device.

It is also important to notice that the proposed design approach is not based on equations but on the optimization of network parameters (currents, voltages, equivalent impedances) at both small and large signals. It suggests a suitable architecture for the realization of active filters based on active inductors. An analytical approach is hard to realize since the filter makes use of AIs that are implemented as closed loop circuits. In addition, it is not useful from a practical point of view also considering that at these frequencies circuits usually make use of distributed elements that are difficult to analytically describe. So the description is strictly related to the network configuration and cannot be generalized.

3. MMIC Design and Test

Following the above proposed design strategy and making use of the new architecture shown in Figure 4 for the AI, a two-cells example filter has been designed, considering for each cell the same architecture shown in Figure 1. However, the core of the proposed work is the AI design and the filter performance are strictly dependent on the active inductor characteristics. The AI should be applied also to different filter families (e.g., Butterworth, Chebyshev, etc.) [11] that make use of grounded inductors. In this example, the filter has been fabricated on a GaAs technology provided by HSRI, realized and tested for demonstration. The standard PDK from HSRI includes 0.13 μm pHEMT devices that exhibit low noise figure, high gain and high-power density (0.7 W/mm). Varactor diodes are not currently available in this technology; the design will be updated with inclusion of varactors as soon as they are available in the future. The proposed design method can be applied with any technology process and into the millimeter-wave band. It is also important to note that the performances of the example filter have been designed for the replacement of an existing passive filter used as post selector in base station unit for mobile communications, and so could be improved further.

The filter acts exactly as a resonator. Its ease of realization represents its effectiveness. More in detail, C_{res} and L_{res} in Figure 1 define the resonant frequency, that is the center frequency of the filter. The quality factor of the filter depends mainly on the quality factors of the components used in the resonator. C_{dc} and L_{dc} realize series resonators that have decoupling effects and are helpful to tighten the filter bandwidth. The same architecture has been widely described in [11–13]. The filter has been realized replying twice the same cell. Each single cell includes an active inductor that realizes L_{res}. The embedded AI in each cell has been realized with a transistor, whose dimensions are 6 × 25 um, in common source configuration. A self-gate bias architecture has also been implemented in order to limit the number of the bias pads and to reduce the temperature dependence of the filter characteristics. The bias current per transistor is 15 mA, while the total power consumption is 120 mW with a DC voltage supply of 4 V. It is straightforward to notice that the DC power consumption is not very low; but in this example this design choice was not critical. In general, it depends on the desired linearity and dynamic range of the filter, in addition to the available technology. Considering the HAMLA13B Model Handbook from HSRI, in particular the recommended bias voltages and the IV-curves of the active devices, a drain voltage of 4 V and a bias current per transistor of 15 mA are suitable choices, considering the required power handling and the connected voltage and current swings. Obviously, by using different technology processes and considering different design specs power consumption should be different, but still in the same order of magnitude if you use the same transistor topology. In addition, it may be reduced also changing the number of cells, so allowing to have a lower shape factor of the filter.

In Figure 6, the complete schematic of the single cell is reported, while in Figure 7, the simulated results of the proposed IC design are shown for a fixed tuning state. The insertion loss is about 6 dB, while the 3 dB bandwidth is 30 MHz. It is worth noting that given the active nature of the filter there is no problem in principle to obtain very low attenuation, close to zero, but this was not the aim of this work since the proposed circuit has been designed for the one-to-one replacement of an existing passive post-selector filter, and therefore it had to have the same performances. In that application the insertion loss is not a critical parameter, and has not been improved. However, active filters may reduce the attenuation to zero or even have amplification, at the expense of stability. Typically, an insertion loss of 0.5 dB can be reached with still a good stability margin. The insertion loss is defined properly balancing the negative resistance introduced by the active inductor and losses of passive components of filtering network. So, it is of primary importance the definition of a suitable shape for the input impedance of the AI that allows to provide low losses, out-of-band rejection and tunability versus an external control voltage. As shown in Figure 7, the shape factor achievable with the proposed solution is typical of higher-order passive solutions [11], and the quality factor is approximately equal to 90. It is important to remark that the circuit is stable, as evident from the behavior of the stability factor K and Delta parameter in the same figure [14]. The stability has been checked not only between the external RF I/O pads, but also at transistor level in each stage with the approach proposed in [8]. This is necessary due to the presence of feedback networks (Figure 4) that may generate inner instability problems. In particular, the amplifier in each active inductor is unconditionally stable at all frequencies. Additionally, each cell is unconditionally stable at all frequencies, because the small negative resistance of the active inductor is compensated by the losses of the resonating capacitor and of the connecting lines. Therefore, no oscillations can take place due to reflections between cells. In Figure 8, the simulated output power and attenuation vs. input power are shown for the circuit tuned at 1800 MHz, for a −7 dBm compression point. In Figure 9, the simulated dynamic load lines of the transistor for the same tuning frequency are also shown for several input power levels, demonstrating class-A operations.

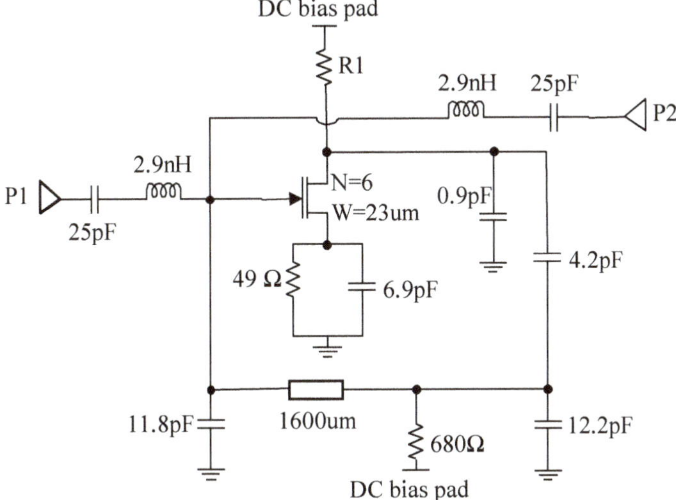

Figure 6. Schematic of the single cell.

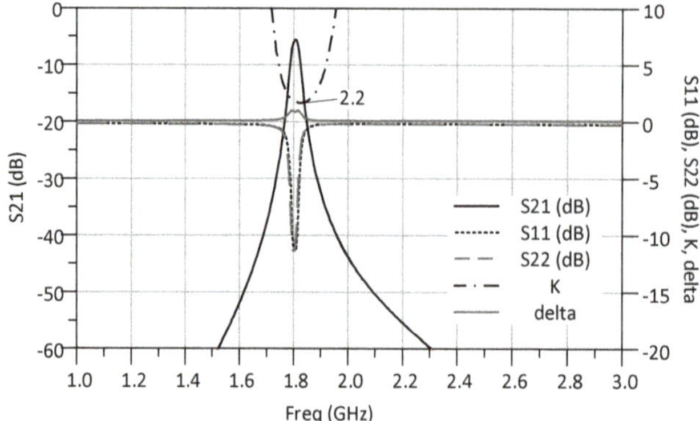

Figure 7. Simulated S-parameters and stability parameters of the two-cell filter for a single frequency.

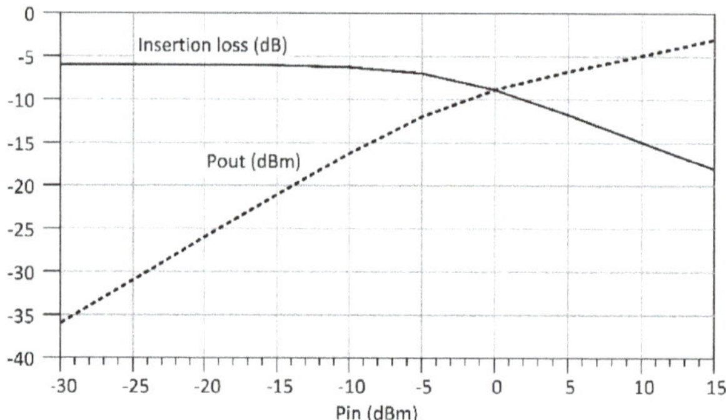

Figure 8. Simulated compression of the filter tuned at 1800 MHz.

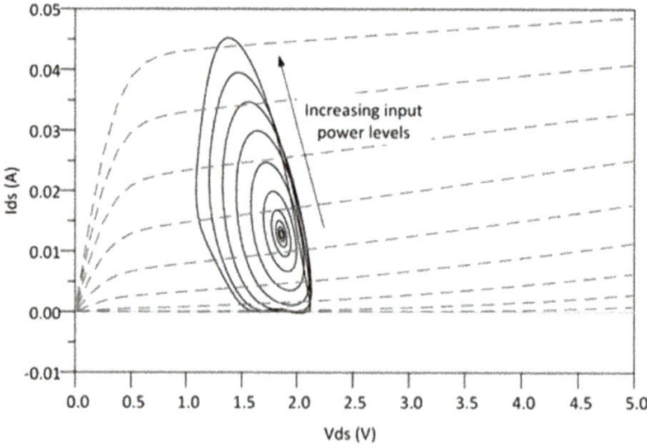

Figure 9. Dynamic load line of the transistor of the filter tuned at 1800 MHz, for different input power levels up to −9 dBm.

At the considered frequencies, better performance can be achieved by using, for instance, SAW devices. However, SAW filters [15] have a limited integration capability and cannot be used at very high frequencies. Additionally, an active chip filter has potential tunability, and smaller size and cost compared to a SAW filter.

A photograph of the fabricated chip is shown in Figure 10a. The chip size is 3 × 1.5 mm, and obviously may be further reduced in next redesign. The two cells are coupled by LC-series elements. The single cell is magnified in Figure 10b, where also some voltages and currents components are reported to better illustrate the proposed approach.

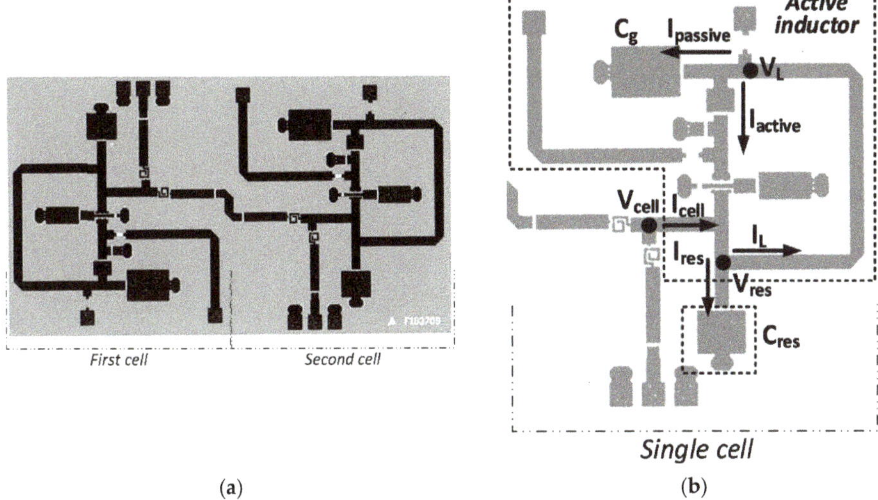

Figure 10. The fabricated active filter: (**a**) Photograph of the chip, (**b**) a single cell where the currents of Figure 4 are indicated.

The behavior of the same electrical parameters is shown in the follows in order to prove the design consideration introduced in the previous Section. The main inductance L in Figure 4 in this example is replaced by a length of line in each cell. Since standard PDK from HSRI does not include varactors, it has not been possible to use variable capacitors, so in this first run they have been replaced by fixed capacitances, for a first assessment of the design approach and in order to avoid further uncertainties that may be introduced by the nonlinear models of the devices. In fact, our aim is to provide a feasibility proof demonstrating the tunability capability of the filter by replacing the unavailable varactors with fixed capacitors. So, three different chips with different capacitance values of the capacitor C_g in Figure 10b have been fabricated, each tuned to a different center frequency. However, the values of the fixed capacitances have been designed with a capacitance ratio of 3:1 for the full tuning bandwidth, making them suitable to be replaced by real varactors. Obviously, there are minor adjustments to be made when the fixed capacitors are replaced by varactors, mainly in term of losses. But we have already verified in different discrete prototypes that the replacement is still possible and does not significantly affect the overall performance of the filter. This is possible thanks to the active nature of the inductors, that can be slightly tuned to compensate the additional losses introduced by varactors.

In Figure 11, comparison between simulated and measured S-parameters for a fixes tuning state is reported demonstrated a good agreement; while measured results of the S-parameters of the three chips, centered at the center frequency and at the two extreme frequencies in the tuning range, are reported in Figure 12. As shown, bandwidth and quality factor are approximately constant at all frequencies. The currents in the active inductor and the corresponding voltages, as indicated in Figure 10b, are plotted in Figures 13–15 vs. time. All quantities refer to the cell tuned at 1800 MHz and at compression level. More in details, in Figure 13 the voltage (V_{cell}) and current (I_{cell}) at the input of the cell are shown. It is apparent that the cell has resistive behavior, but very low losses, as indicated by the very low current. In Figure 14 the currents (I_{res}) in the resonating capacitance and in the active inductance (I_L) are shown, together with the voltage at that node (V_{res}). The sum of the two currents (I_{res} and I_L) is almost zero since they are in anti-phase, indicating their almost total cancellation. In Figure 15, the inductor current I_L, the current $I_{passive}$ through the capacitor C_g and the current I_{active} from the transistor are shown, together with the voltage at the corresponding node. As stated in the previous section, it is apparent that also in this example the main contribution of the inductor current flows through the capacitor C_g to ground, and that it is a passive current, as indicated by the phase relation between voltage and current.

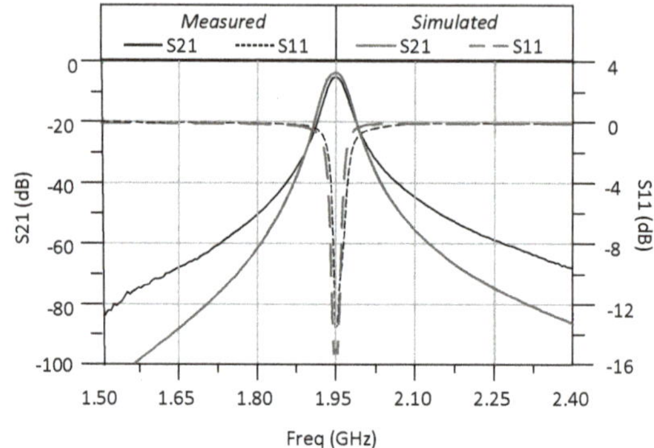

Figure 11. Comparison between simulated and measured S-parameters at center frequency.

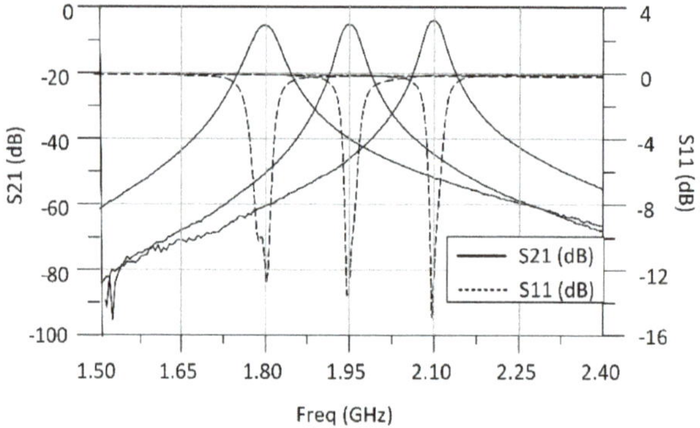

Figure 12. Measured S-parameters of the two-cell filter at three frequencies in the tuning range.

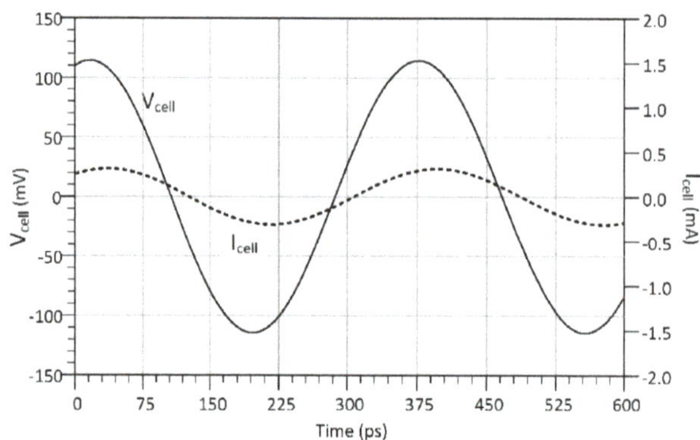

Figure 13. Simulated voltage and current at the input of the resonating cell (see Figure 10b).

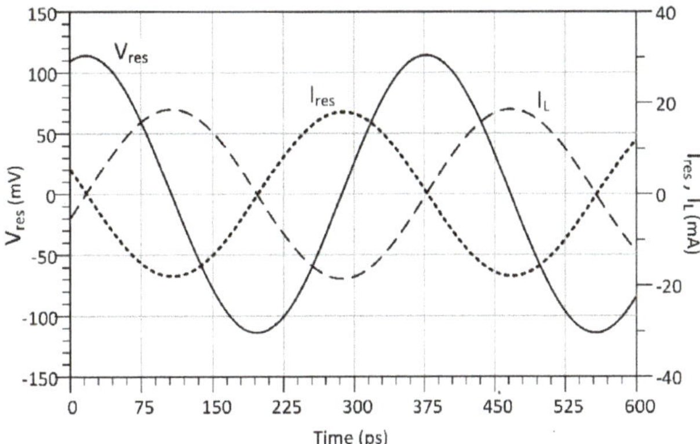

Figure 14. Simulated currents I_{RES} in the resonating capacitance C_{RES} and in the active inductor I_L, together with the voltage V_{RES} at the corresponding node (see Figure 10b).

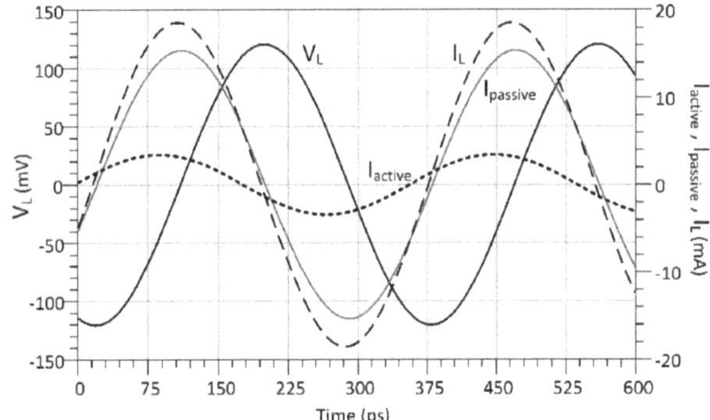

Figure 15. Simulated voltage and current at the input of the resonating cell (see Figure 10b).

A minor contribution comes from the transistor, and it is an active current. Their combination is a slightly active current, corresponding to a negative equivalent resistance, that compensates for the losses in the resonator capacitance (C_{res}). Thus, these real waveforms illustrate and confirm the design approach represented in Figures 4 and 5, that minimizes the RF current in the transistor and; therefore, the bias current requirement for the transistor itself.

The noise figure of the filter has not been measured, since it was not optimized in this design. Simulations give a noise figure in the order of 12 dB. However, reduction techniques [6] should reduce it to approximately 8 dB, approaching the noise figure of a filter with similar losses. Noise reduction techniques will be specifically addressed in a successive design.

It is also important to notice that in the proposed design example of Figure 6, passive components are optimized for the considered frequency and they change accordingly to the operational bandwidth. In any case, as showed in the layout photograph of Figure 10 the area waste due to both the number of components and their values is minimal; in addition some passive components can be also replaced by distributed elements moving at higher frequencies, so further reducing the chip area. In general, the same networks definition may change moving the design to different frequencies, since it strictly depend also on the technology and specs, but still preserving around the same area occupation and realizing the same design principle illustrated in Figure 4. The proposed design is also robust towards tolerances and spread effects. It has been verified also measuring different chips and all of them show similar performance.

So the design choices allow to obtain a good reliability making the filter a good candidate to be used in practical applications.

Finally, for completeness, in Table 1 a summary of measurement results is also presented. Both linear and nonlinear characteristics are in good agreement with simulations, showing the feasibility and reliability of the proposed I_C design. Table 2 shows also a comparison with different literary solutions. From Table 2 it is evident that this work has the highest power consumption, but it is important to notice that the power consumption mainly depends on the selected technology. In the same table, almost all the cited works are realized with CMOS technologies; while just one work makes use of a p-HEMT technology. Obviously, the power consumption of MOS technologies is significantly lower; but on the opposite this kind of technologies cannot provide proficient results at higher frequencies, mainly in term of power handling capability and tunability. The power consumption of the reference makes use of p-HEMT technology around one half (so on the same order of magnitude) of that shown by our work, but it has also a $P_{-1\,dB}$ compression point significantly lower. A higher power consumption is usually necessary to obtain better power handling capability. Our work is not strictly focused on performance, but it should be considered as a proof-of-concept and demonstrates the reliability of the proposed design approach. By using HEMT devices, the same design scheme should be successfully applied at also tens of GHz, ensuring good tunability and high 1 dB power compression point.

Table 1. Summary of measured results.

Frequency f_0 (GHz)	IL (dB)	$BW_{-3\,dB}$ (MHz)	$P_{-1\,dB}$ (dBm)	OIP3 (dBm)
2.102	5.62	26	−8.60	0.5
1.950	6.28	29	−9.54	−0.6
1.785	7.84	33	−9.31	−0.5

Table 2. Comparison with different literary solutions.

References	Topology	Center Frequency [MHz]	$BW_{3\,dB}$ [MHz]	Tuning Range [%]	$P_{1\,dB}$ [dBm]	Power Consumption [mW]
[16]	BiCMOS (0.13 um)	3375	-	66	−8	96
[17]	CMOS (0.13 um)	2000	-	66	−6	38
[18]	CMOS (45 nm)	2511	36	-	−1.5	0.18
[19]	CMOS (40 nm)	140	5÷56	100	12	38
[20]	CMOS (0.18 um)	5.3×10^3	1.7×10^3	-	2.5	2.2
[21]	p-HEMT (0.15 um)	22.6×10^3	900	8.8	−20	50
[22]	BiCMOS (0.13 um)	53.85×10^3	14×10^3	-	−3.67	2.62
This work	**p-HEMT (0.13 um)** *(Two-cell filter)*	**1950**	**29**	**15**	**−8.5**	**120**

4. Conclusions

We presented an innovative design approach for the minimization of the bias current in the AI-based active filters. The design strategy here proposed allows to obtain a significant reduction of the required DC power, without significantly affecting the power handling capability. The core of the solution here addressed for the first time is the active inductor design. Both the circuitry architecture and the components values were optimized to this purpose and the design choices were described in detail. We also showed how the equivalent inductance and resistance can be tuned by means of variable capacitances. An example of application was also provided: An integrated active filter, centered at three different frequencies, was realized with pHMET technology provided by HSRI Foundry. The filter was optimized between 1800 and 2100 MHz with a narrow and almost constant 3 dB bandwidth of about 30 MHz and a high shape factor, typical of higher order passive filters. This solution allows one to obtain a dynamic range of about 75 dB, minimizing the power consumption. All these characteristics, as well as measured results, makes the proposed solution a good candidate to be used in practical applications and many RF TX/RX systems.

Author Contributions: Conceptualization, G.L., V.S. and L.P.; investigation, L.P. and V.S.; data curation, H.L. and Z.H.; supervision, G.L.; validation, H.L.; writing and editing, L.P., G.L. and V.S. All authors have read and agreed to the published version of the manuscript.

Funding: This research received no external funding.

Conflicts of Interest: The authors declare no conflict of interest.

References

1. Aghazadeh, S.; Martinez, H.; Saberkari, A. 5 GHz CMOS All-Pass Filter-Based True Time Delay Cell. *Electronics* **2018**, *8*, 16. [CrossRef]

2. Safari, L.; Barile, G.; Ferri, G.; Stornelli, V. High performance voltage output filter realizations using second generation voltage conveyor. *Int. J. RF Microw. Comput. Aided Eng.* **2018**, *28*, e21534. [CrossRef]

3. Ben Hammadi, A.; Haddad, F.; Mhiri, M.; Saad, S.; Besbes, K. RF and Microwave Reconfigurable Bandpass Filter Design Using Optimized Active Inductor Circuit. *Int. J. RF Microw. Comput. Aided Eng.* **2018**, *28*, e21550. [CrossRef]

4. Thanachayanont, A.; Sae Ngow, S. Class AB VHF CMOS active inductor. In Proceedings of the 2002 45th Midwest Symposium on Circuits and Systems, 2002, MWSCAS-2002, Tulsa, OK, USA, 4–7 August 2002.

5. Pantoli, L.; Stornelli, V.; Leuzzi, G. Tunable active filters for RF and microwave applications. *J. Circuits Syst. Comput.* **2014**, *23*, 1450088. [CrossRef]

6. Pantoli, L.; Stornelli, V.; Leuzzi, G.; Hongjun, L.; Zhifu, H. GaAs MMIC tunable active filter. In Proceedings of the 2017 Integrated Nonlinear Microwave and Millimetre-wave Circuits Workshop (INMMiC), Graz, Austria, 20–21 April 2017.

7. Pantoli, L.; Stornelli, V.; Leuzzi, G. Low-noise tunable filter design by means of active components. *Electron. Lett.* **2016**, *52*, 86–88. [CrossRef]

8. Pantoli, L.; Stornelli, V.; Leuzzi, G. A low-voltage low-power 0.25 µm integrated single transistor active inductor-based filter. *Analog Integr. Circuits Signal Process.* **2016**, *87*, 463–469. [CrossRef]

9. Foster, R. A Reactance Theorem. *Bell Syst. Tech. J.* **1924**, *3*, 259–267. [CrossRef]

10. Merrill, J. Theory of the Negative Impedance Converter. *Bell Syst. Tech. J.* **1951**, *30*, 88–109. [CrossRef]

11. Zverev, A.I. *Handbook of Filter Synthesis*; John Wiley & Sons: New York, NY, USA, 1967.

12. Rhea, R.W. *Filter Synthesis Using Genesys S/Filter*; Artech House Publishers: Boston, MA, USA, 2014.

13. Matthaei, G.L.; Young, L.; Jones, E.M. *Microwave Filters. Impedance Matching Networks and Coupling Structures*; Artech House Books: Boston, MA, USA, 1980.

14. Platzker, A.; Struble, W.; Hetzler, K.T. Instabilities diagnosis and the role of K in microwave circuits. In Proceedings of the IEEE MTT-S International Microwave Symposium Digest, Atlanta, GA, USA, 14–18 June 1993; Volume 3, pp. 1185–1188.

15. Lu, X.; Mouthaan, K.; Soon, Y.T. Wideband Bandpass Filters With SAW-Filter-Like Selectivity Using Chip SAW Resonators. *IEEE Trans. Microwave Theory Tech.* **2014**, *62*, 28–36. [CrossRef]

16. Mohammadi, L.; Koh, K. 2–4 GHz Q-tunable LC bandpass filter with 172-dBHz peak dynamic range, resilient to +15 dBm out-of-band blocker. In Proceedings of the 2015 IEEE Custom Integrated Circuits Conference (CICC), San Jose, CA, USA, 28–30 September 2015.

17. Mohammadi, L.; Koh, K. Low power highly linear band-pass/band-stop filter for 2–4 GHz with less than 1% of fractional bandwidth in 0.13 µm CMOS technology. In Proceedings of the 2017 IEEE Radio Frequency Integrated Circuits Symposium (RFIC), Honolulu, HI, USA, 4–6 June 2017.

18. Kumar, V.; Mehra, R.; Islam, A. A 2.5 GHz Low Power, High-Q, Reliable Design of Active Bandpass Filter. *IEEE Trans. Device Mater. Reliab.* **2017**, *17*, 229–244. [CrossRef]

19. Wu, B.; Chiu, Y. A 40 nm CMOS Derivative-Free IF Active-RC BPF With Programmable Bandwidth and Center Frequency Achieving Over 30 dBm IIP3. *IEEE J. Solid-State Circuits* **2015**, *50*, 1772–1784. [CrossRef]

20. Wang, S.; Lin, W. C-band complementary metal-oxide-semiconductor bandpass filter using active capacitance circuit. *IET Microw. Antennas Propagat.* **2014**, *8*, 1416–1422. [CrossRef]

21. Fan, K.W.; Weng, C.C.; Tsai, Z.M.; Wang, H.; Jeng, S.K. K-band MMIC active band-pass filters. *IEEE Microw. Wirel. Compon. Lett.* **2005**, *15*, 19–21.

22. Chaturvedi, S.; Božanić, M.; Sinha, S. 60 GHz BiCMOS active bandpass filters. *Microelectron. J.* **2019**, *90*, 315–322. [CrossRef]

 *electronics*

Article

Sinusoidal Oscillators Operating at Frequencies Exceeding Unity-Gain Bandwidth of Operational Amplifiers

Vladimir Ulansky [1,*] and Ahmed Raza [2]

[1] Department of Electronics, Robotics, Monitoring and IoT Technologies, National Aviation University, 03058 Kyiv, Ukraine

[2] Projects and Maintenance Section, The Private Department of the President of the United Arab Emirates, Abu Dhabi 000372, UAE; ahmed.awan@dopa.ae

* Correspondence: vladimir_ulansky@nau.edu.ua

Received: 26 April 2020; Accepted: 17 May 2020; Published: 20 May 2020

Abstract: This paper proposes a novel operational amplifier (OPA) voltage-controlled oscillator (VCO) circuits on the basis of impedance converters. The VCO can operate over a frequency range exceeding unity-gain bandwidth due to the location of the tank circuit, not at the output of the OPA, but at the noninverting input. The paper presents the mathematical modeling of oscillated amplitude and start-up conditions. The simulation results confirm the theoretical achievements. The designed and simulated VCO uses an ultra-low noise wideband OPA LMH6629MF, covers a frequency band between 0.830 GHz and 1.429 GHz, and exhibits a maximum in-band total harmonic distortion (THD) of 1.7%. It has a maximum in-band phase noise of −139.3 dBc/Hz at 100 kHz offset frequency and has an outstanding value of a standard figure of merit (FoM) of −198.6 dBc/Hz. The zero-peak amplitude of output voltage is from 3.2 V to 4 V for all generated frequencies at a supply voltage of ±5 V. The fabricated prototype-oscillator based on OPA LMH6624 operates at a frequency of 583.1 MHz with a power level of 0 dBm.

Keywords: operational amplifier; voltage-controlled oscillator; unity-gain bandwidth; varactor; total harmonic distortion; phase noise

---

1. Introduction

Sinusoidal oscillators are used in almost all systems of receiving and transmitting information, as well as in measuring instruments and systems. One of the most rapidly developing class of sinusoidal oscillators is the class of the voltage-controlled oscillator (VCO). Voltage-controlled oscillators are commonly used in digital frequency synthesizers, which are one of the main subsystems of modern communications systems. The rapid development of modern communications and instrumentation systems has created a high demand for low noise VCOs [1–3]. Modern microwaves sinusoidal oscillators use bipolar junction transistors (BJT), heterojunction bipolar transistors (HBT), a low-noise high-electron-mobility-transistors (HEMT) and pseudomorphic HEMT (pHEMT) as active devices for achieving low phase-noise performance [4–6]. The appearance on the market of a ultra-low-noise, high-speed, broadband operational amplifier (OPA) creates the possibility of their use in the sinusoidal oscillators in the ultra-high frequency range. For example, the LMH6629MF (Texas Instruments) OPA has an input noise voltage of 0.69 nV/√Hz at the corner frequency of 4 kHz, a slew rate of 1600 V/μs and small-signal −3 dB bandwidth of 900 MHz [7]. Low-noise, high-speed OPA oscillators can offer a practical alternative to transistor oscillators, the performance of which to some extent depends on the variability of the transistor small-signal parameters. The use of a low-noise, high-speed OPA as an active device in the oscillator circuits has some advantages [8,9]. An operational amplifier is an amplifier with very high input impedance and very low output impedance; it is easy to introduce required positive feedback around the OPA; the oscillator design is simple due to the lack of the bias circuit; the oscillator circuit does not require any adjustment during fabrication.

As is well known [8,10,11], oscillators based on the Colpitts and Hartley topology require a relatively high voltage gain to start-up oscillation. High voltage gains reduce the maximum frequency of oscillation because of the limited gain-bandwidth product of OPA. Besides, the maximum frequency of generated sinusoidal oscillations at the output of the OPA in the Colpitts and Hartley topologies is limited by the ratio of the slew rate to the amplitude of voltage oscillations multiplied by two pi [12]. Thus, with the required voltage amplitude of, say, three volts, and the use of the LMH6629 OPA, the maximum achievable frequency will be less than 85 MHz, which is very far from the microwave frequency range. By simple calculations, we can estimate what should be the slew rate of an OPA to achieve a frequency of 1 GHz with a 3 V amplitude of oscillations. It should be about 19,000 V/μs. Currently, such a slew rate in OPA is unattainable.

The VCO topology proposed in [13] uses the OPA circuit with negative input inductance observed at the noninverting input of the OPA. The disadvantage of this circuit is the use of two resistors, one of which presents the positive feedback circuit, and the second connects the inverting input of the OPA to the ground. These resistors are the source of thermal noise.

In this study, we propose several new VCO topologies that use the idea of a negative impedance converter. However, unlike the well-known studies [14–17], the converter circuit does not include resistors. In contrary to the Colpitts and Hartley oscillators requiring a sufficiently high voltage gain for self-excitation, the proposed oscillator circuits operate with a voltage gain of less than unity; this feature significantly increases the operating frequency, which can exceed the unity-gain bandwidth of the OPA. This particularity of the proposed VCO circuits is due to the location of the tank circuit not at the output of the OPA, but at the noninverting input. Mathematical modeling, simulation, and prototype implementation of the proposed oscillators are given.

2. Architecture of Oscillators

Figure 1 shows the general VCO electronic diagram. The circuit inside the dashed rectangular is the negative impedance converter. The input impedance observed at the noninverting terminal of the OPA is as follows [16]:

$$Z_{in} = -Z_2 \frac{Z_1}{Z_0} \tag{1}$$

where Z_{in} is the input impedance seen by the noninverting terminal of the OPA, Z_0 is the impedance between the inverting input and output of the OPA, Z_1 is the impedance between the noninverting input and output of the OPA, and Z_2 is the impedance between the inverting terminal of the OPA and ground. The step-by-step derivation of Equation (1) is given in [18] (pp. 349–350).

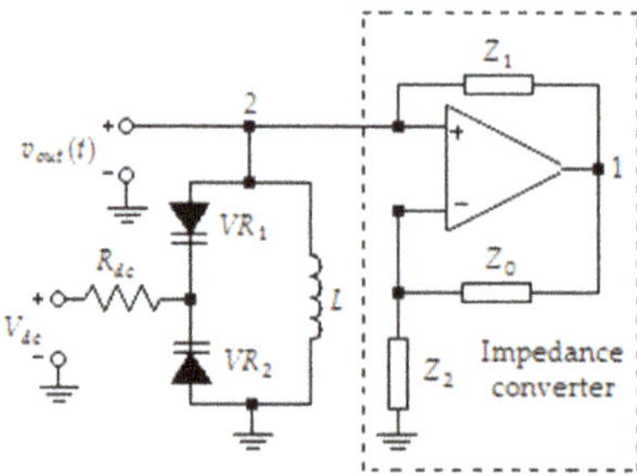

Figure 1. General electronic diagram of voltage-controlled oscillator on the basis of an impedance converter.

In Reference [15–17], one of the impedances Z_0, Z_1, or Z_2 is capacitive, and the other two are resistive.

The presence of significant resistances in the circuit of any oscillator leads to an increase in thermal noise [19]; therefore, in this study, the impedances Z_0, Z_1, and Z_2 are capacitive or inductive.

In the circuit of Figure 1, inductor L and two contrary connected varactors VR_1 and VR_2 present the tank circuit of the VCO. Resistor R_{dc} isolates the dc control voltage line from the VCO tank.

As can be seen in Figure 1, the tank circuit is connected to the noninverting input of the OPA rather than its output. This feature of the proposed VCO allows extending operation frequency range beyond the unity-gain bandwidth of OPA.

Figure 2 shows VCO circuits based on impedance converters with two inductors and one capacitor. The circuits in Figure 2a,b introduce positive feedback through inductor L_1 and capacitor C_1, respectively.

Figure 3 presents VCO circuits on the base of impedance converters with three inductors (a) and two capacitors and one inductor (b).

The oscillator circuits based on impedance converters in which a capacitor is used between the inverting input of the OPA and ground are not considered.

Further, in the article, we will analyze in detail the VCO circuit presented in Figure 2a, and where necessary, we will refer to other VCO configurations.

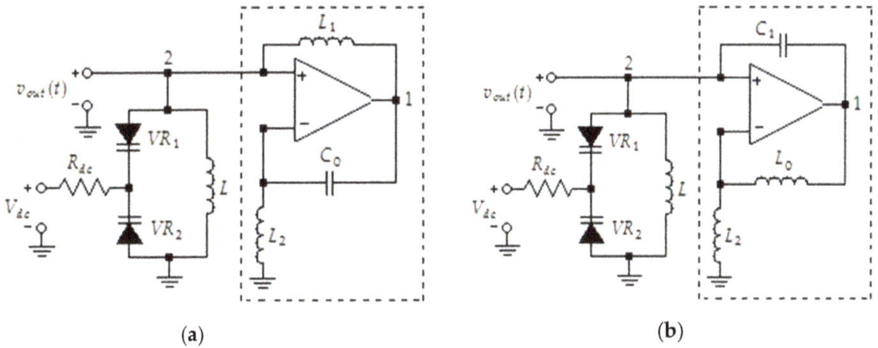

(a)   (b)

Figure 2. VCO circuits based on impedance converters with two inductors and one capacitor; (a) with inductive positive feedback and (b) with capacitive positive feedback.

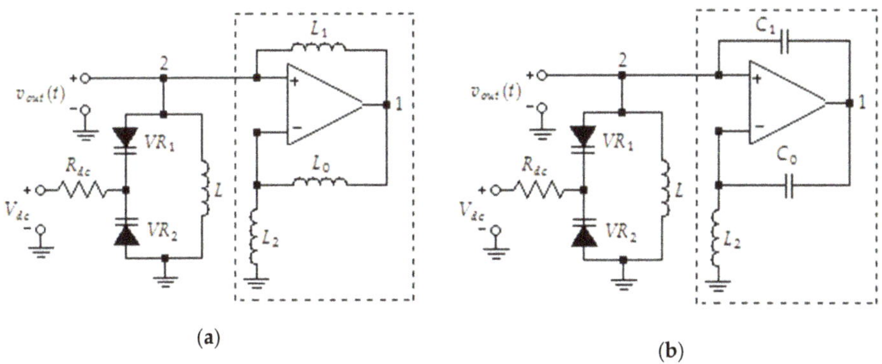

(a)   (b)

Figure 3. Voltage-controlled oscillator (VCO) circuits based on the impedance converter with three inductors (**a**) and one inductor and two capacitors (**b**).

3. Converter Analysis

Before analyzing the VCO circuit of Figure 2a, let's examine the operation of the converter at frequencies exceeding the OPA bandwidth. Figure 4a shows the test circuit. We assume that the OPA unity-gain bandwidth is 900 MHz, the slew rate is 1600 V/µs, the open-loop gain is 1000 V/V, and the test signal is a sinusoidal voltage with a frequency of 1 GHz and peak amplitude of 2 V.

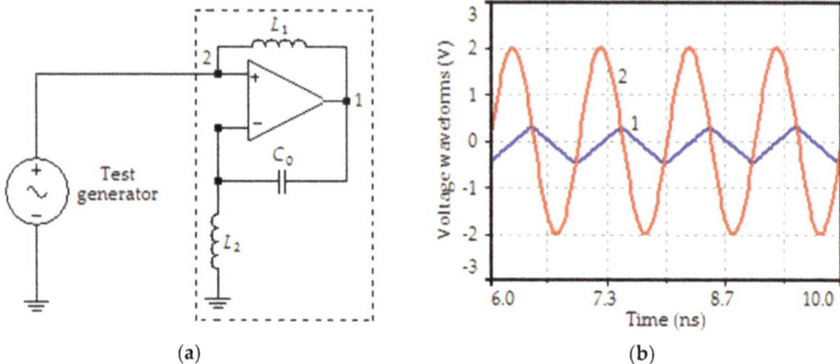

Figure 4. (**a**) Impedance converter testing scheme; (**b**) Input (curve 2) and output (curve 1) voltages of the test circuit.

Figure 4b shows the voltage waveforms at the noninverting input (curve 2) and the output of the OPA (curve 1). As we can see in Figure 4b, the slew rate of the OPA is not sufficient to ensure the shape of the OPA output voltage as that at the noninverting input. Therefore, we observe a triangular voltage waveform at the OPA output (node 1). Moreover, Figure 4b shows that the OPA output voltage lags the input voltage by 90°. From a comparison of the circuits in Figures 2a and 4a, we can see that instead of the test generator in Figure 4a, a parallel resonant circuit is used in Figure 2a. We will show further that the shape of the voltage curves in the circuit of Figure 2a will be the same as in Figure 4b; however, curve 2 will correspond to the voltage at the output of the VCO.

4. VCO Analysis

The electronic part of the VCO introduces an alternating current (ac) source $i_{OPA}(t)$ into the tank circuit, as shown in Figure 5, where C_{VCO} and L_{VCO} are, respectively, the VCO total capacitance and inductance, and R_{par} is the equivalent parallel resistance of the tank circuit at the fundamental frequency. Essentially the current $i_{OPA}(t)$ is the feedback current flowing through inductor L_1.

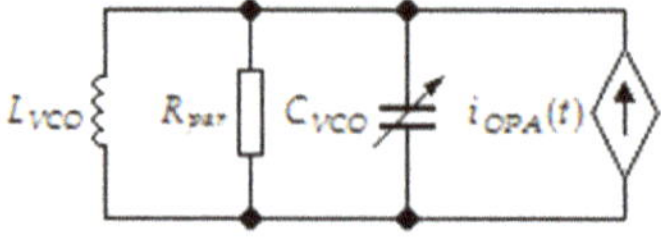

Figure 5. AC equivalent circuit of the VCO.

The VCO total capacitance includes the capacitance of the contrary connected varactors VR_1 and VR_2, the input capacitance of the OPA, and the parasitic capacitance of the printed circuit board (PCB). Therefore,

$$C_{VCO}(V_{dc}) = \frac{C_{VR1}(V_{dc})C_{VR2}(V_{dc})}{C_{VR1}(V_{dc}) + C_{VR2}(V_{dc})} + C_{PCB} + C_{OPA} \tag{2}$$

As we can see in Equation (2), the VCO capacitance is a function of voltage V_{dc} because the varactor capacitance depends on this voltage. Onwards, inductor L_1 provides positive shunt–shunt feedback.

By applying the y-parameter analysis to the positive feedback network, we find that y_{11} parameter incorporates into the tank circuit. Since $y_{11} = 1/j\omega_{fun}L_1$, inductance L_1 appears in parallel with tank inductance L, where ω_{fun} is the fundamental angular frequency of oscillation.

Therefore, the total VCO tank circuit inductance is given by

$$L_{VCO} = L_1 \| L \tag{3}$$

The current $i_{OPA}(t)$ flows through the tank circuit. However, only the first harmonic of the current flow creates a significant voltage drop across the tank. The second, third, and subsequent current harmonics create insignificant voltage drops that can be neglected.

We should also note that the tank circuit is connected to the noninverting input of the OPA, where the input impedance is extremely high. Therefore, the OPA does not practically short-out the tank circuit.

This property of the proposed VCO allows obtaining high voltage amplitude of the generated voltage $v_{out}(t)$, which is not limited by the OPA output. Indeed, we can describe the VCO output voltage by the following equation:

$$v_{out}(t) = \left|I_{OPA,1}(V_{dc})\right| R_{par}(V_{dc}) \sin\left[\omega_{fun}(V_{dc})t + \varphi_1(V_{dc})\right] \tag{4}$$

where $|I_{OPA,1}(V_{dc})|$, $\omega_{fun}(V_{dc})$, and $\varphi_1(V_{dc})$ are, respectively, the amplitude, frequency and initial phase of the first harmonic of current $i_{OPA}(t)$. Therefore, we can see from (4) that the amplitude of the generated sinusoidal voltage is proportional to $|I_{OPA,1}|$ and R_{par}. In a virtual case of lossless tank circuit, i.e., when $R_{par} \rightarrow \infty$, the amplitude of the output voltage would tend to infinity.

This is a unique property of the proposed OPA oscillator because, for the Colpitts and Hartley OPA oscillators [8], the output amplitude satisfies the following inequality:

$$V_m \le \frac{4V_{sat}R_{par}}{\pi(R_C + R_{par})} \tag{5}$$

where V_{sat} is the OPA saturation voltage, and R_C is the resistance of the load resistor connected between the OPA output and tank circuit.

Since the OPA saturation voltage in (5) is less than the power supply voltage V_{cc}, the oscillation amplitude is also less. Figure 6a illustrates the behavior of the OPA output voltage (curve 1) and the oscillator output voltage (curve 2) for the Colpitts oscillator operating at frequency of 53 MHz with $V_{cc} = \pm 5$ V and voltage peak amplitude of 3.7 V. We can also observe in Figure 6a that the shape of the voltage at the output of the OPA, in this particular case, is trapezoidal, which means the slew rate of the OPA is not high enough to ensure a rectangular shape.

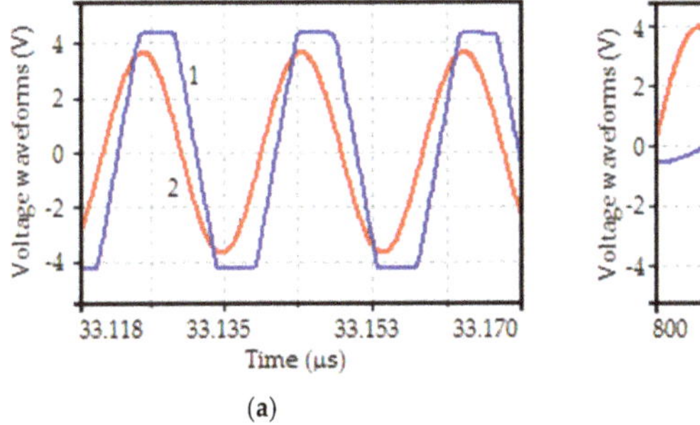
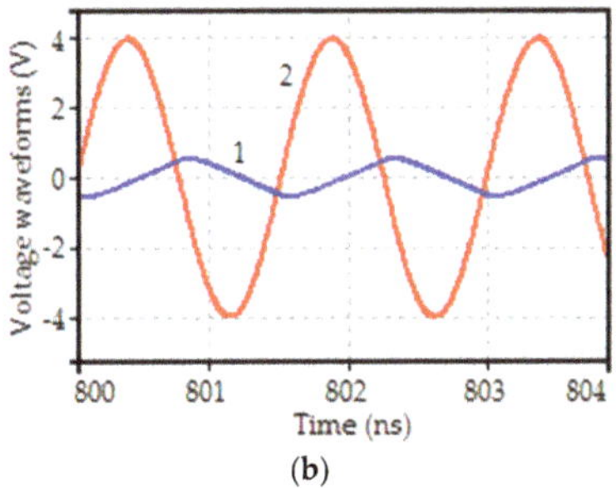

Figure 6. (a) OPA output voltage (curve 1) and oscillator output voltage (curve 2) for the Colpitts VCO; (b) OPA output voltage (curve 1) and oscillator output voltage (curve 2) for the proposed VCO.

With a further increase in the frequency of the generated oscillations in the Colpitts oscillator, the voltage shape at the output of the OPA becomes triangular with the amplitude less than V_{sat}. In this case, also, the amplitude of sinusoidal oscillations in the Colpitts oscillator will always be less than the magnitude of triangular oscillations.

A completely different relationship exists between the amplitudes of the voltages at the output of the OPA (node 1) and the oscillator (node 2) in the circuit in Figure 2a. Figure 6b illustrates the behavior of the OPA output voltage (curve 1) at node 1 and the oscillator output voltage (curve 2) at node 2 for the proposed oscillator operating at frequency of 667 MHz with the same OPA, power supply voltages and output peak amplitude of 4 V. As can be seen in Figure 6b, the amplitude of the sinusoidal voltage can be much larger than the voltage amplitude at the output of the OPA (node 1).

Therefore, the proposed VCO can operate at frequencies significantly exceeding the unity-gain bandwidth of OPA.

It should also be noted that by choosing a low-noise OPA, the phase noise of the VCO will be determined mainly by the noise of the tank circuit.

5. Amplitude of Oscillations

Let us determine the amplitude of the sinusoidal voltage at the output of the VCO (node 2). Assume the voltage at node 1 be triangular, as shown in Figure 6b (curve 1). For the triangle wave, we can describe the voltage at node 1 by the complex exponential Fourier series as follows:

$$v_{OPA}(t) = 0.5 \sum_{\mu=-\infty}^{\infty} V_{\mu} e^{j\mu\omega_{fun}t} \tag{6}$$

where V_{μ} is the complex amplitude of the harmonic number μ of voltage $v_{OPA}(t)$.

The complex amplitude V_{μ} can be represented as

$$V_{\mu} = |V_{\mu}| e^{j\theta_{\mu}} \tag{7}$$

where $|V_{\mu}|$ and θ_{μ} are the amplitude and phase of the voltage harmonic number μ, respectively.

As is well known [20], the amplitude of the voltage harmonic μ for the triangle wave is

$$|V_{\mu}| = \frac{8V_{triangle}}{\pi^2\mu^2} \sin\frac{\mu\pi}{2} \tag{8}$$

where $V_{triangle}$ is the amplitude of the triangular voltage at the OPA output.

The feedback current $i_{OPA}(t)$, which flows from node 1 into the tank circuit, we also present by the complex exponential Fourier series

$$i_{OPA}(t) = 0.5 \sum_{\mu=-\infty}^{\infty} I_{OPA,\mu} e^{j\mu\omega_{fun}t} \tag{9}$$

where $I_{OPA,\mu}$ is the complex amplitude of the current harmonic number μ.

The complex amplitude $I_{OPA,\mu}$ we write in polar form as follows:

$$I_{OPA,\mu} = |I_{OPA,\mu}| e^{j\varphi_{\mu}} \tag{10}$$

where $|I_{OPA,\mu}|$ and φ_{μ} are the amplitude and phase of the current harmonic number μ, respectively.

In the ac equivalent circuit of the VCO shown in Figure 5, the feedback inductor L_1 is in parallel with tank inductor L. The feedback current $i_{OPA}(t)$ flows through inductor L_1 and tank circuit due to the applied voltage $v_{OPA}(t)$. Considering the forward transfer admittance of the feedback network

$y_{12} = -1/j\omega_{fun}L_1$ and parallel connection of inductors L_1 and L in the equivalent circuit of Figure 5, by Ohm's law, we have

$$I_{OPA,\mu} = -V_\mu / \left(j\omega_{fun}L_1 \| L \right) = j V_\mu / \left(\omega_{fun}L_1 \| L \right) \tag{11}$$

By substitution of (7) and (8) into (11), we get

$$I_{OPA,\mu} = \frac{8 V_{triangle}}{\omega_{fun}L_1 \| L \pi^2 \mu^2} \sin\frac{\mu\pi}{2} \exp\left[j\left(\theta_\mu + \frac{\pi}{2} \right) \right] \tag{12}$$

The first harmonic of current $i_{OPA}(t)$ is of interest because for the higher harmonics, the equivalent resistance of the tank circuit is negligible, and they do not create a significant voltage drop. By substitution $\mu = 1$ into (12) gives

$$I_{OPA,1} = \frac{8 V_{triangle}}{\omega_{fun}L_1 \| L \pi^2} \exp\left[j\left(\theta_1 + \frac{\pi}{2} \right) \right] \tag{13}$$

We determine the first harmonic of the voltage across the tank by multiplying the complex amplitude $I_{OPA,1}$ and the equivalent resistance of the tank at resonance R_{par} as follows:

$$V_{out,1} = \frac{8 V_{triangle} R_{par}}{\omega_{fun}L_1 \| L \pi^2} \exp\left[j\left(\theta_1 + \frac{\pi}{2} \right) \right] \tag{14}$$

By comparing (7) when $\mu = 1$ and (14), we can state that the first harmonic of voltage across the tank leads the first harmonic of triangular voltage at the OPA output by 90°, which is confirmed by Figure 6b.

From (14) it follows that the amplitude of the voltage across the tank is

$$\left| V_{out,1} \right| = \frac{8 V_{triangle} R_{par}}{\omega_{fun}L_1 \| L \pi^2} \tag{15}$$

Analyzing (15), we can observe that the amplitude of the VCO output voltage $|V_{out,1}|$ is directly proportional to resistance R_{par} and inversely proportional to inductance $L_1 \| L$. As already indicated in the analysis of Equation (4), when using a tank circuit with low losses, the voltage amplitude at the output of the VCO can be significant.

Let us simplify (15) by considering the case when $L_1 \gg L$. In this case, $L_1 \| L \approx L$ and we can write Equation (15) in the following form:

$$\left| V_{out,1} \right| = \frac{8 V_{triangle} R_{par}}{\omega_{fun} L \pi^2} \tag{16}$$

At resonance we have

$$\omega_{fun} L = \sqrt{\frac{L}{C_{VCO}}} = \rho \tag{17}$$

where ρ is the tank circuit characteristic impedance.

As is well known [21] (p. 909), the equivalent resistance of the parallel tank at resonance is

$$R_{par} = \frac{\rho^2}{r_s} \tag{18}$$

where r_s is the series loss resistance of the tank circuit.

Substituting (17) and (18) into (16) gives

$$\left|V_{out,1}\right| = \frac{8V_{triangle}\rho}{r_s\pi^2} \tag{19}$$

Since the ratio of ρ to r_s in (19) is equal to the quality factor of the parallel tank circuit (Q), then we write (19) in the following form:

$$\left|V_{out,1}\right| = \frac{8V_{triangle}Q}{\pi^2} \tag{20}$$

We derived Equation (20) under the condition of an ideal OPA with infinite input resistance, which does not load the tank circuit. However, in the real OPA, the input resistance is not infinite; therefore, in (20), we should use the loaded quality factor. Therefore,

$$\left|V_{out,1}\right| = \frac{8V_{triangle}Q_L}{\pi^2} \tag{21}$$

where Q_L is the quality factor of the loaded VCO tank circuit.

For the VCO shown in Figure 2b, following the same analysis as for the circuit of Figure 2a, we obtain that the amplitude of oscillations is

$$\left|V_{out,1}\right| = \frac{8V_{triangle}\omega_f C_{VCO}\|C_1 R_{par}}{\pi^2} = \frac{8V_{triangle}\omega_f(C_{VCO}+C_1)R_{par}}{\pi^2} \tag{22}$$

From the analysis of Equation (22), it follows that the capacitance of the positive feedback capacitor C_1 should be much less than the capacitance C_{VCO}; in this case, the capacitance C_1 will not affect the frequency of generated oscillations.

Given that at the resonant frequency

$$\omega_f(C_{VCO}+C_1) = \frac{1}{\rho} \tag{23}$$

We transform Equation (22) as follows:

$$\left|V_{out,1}\right| = \frac{8V_{triangle}R_{par}}{\pi^2\rho} \tag{24}$$

Considering that $R_{par}/\rho = Q$ in (24), we obtain again Equation (20), which transforms to (21) due to the loaded tank circuit.

The amplitude of the generated sinusoidal voltages in the VCO circuits in Figure 3a,b is also calculated by Equation (21).

6. VCO Start-Up Conditions

According to the Barkhausen criteria [22], to provide the sustained oscillations, the following conditions must have a place:

$$|A\|\beta| = 1 \tag{25}$$

$$\varphi_A + \varphi_\beta = 0 \tag{26}$$

where $|A|$ and $|\beta|$, and φ_A and φ_β are, respectively, the gains and initial phases of impedance converter and feedback network.

As shown in Figure 4b, the voltage gain of the impedance converter is less than unity. We determine the feedback network ratio as follows:

$$|\beta| = \frac{\left|V_{out,1}\right|}{|V_1|} \tag{27}$$

where $|V_1|$ is the voltage amplitude of the first harmonic of triangular voltage at the output of the OPA (node 1).

Substituting (8) at $\mu = 1$ and (21) into (27) gives

$$|\beta| = Q_L \tag{28}$$

We can conclude from (25) and (28) that at the steady-state the gain of the impedance converter is

$$|A| = \frac{1}{|\beta|} = \frac{1}{Q_L} \tag{29}$$

To start the oscillations, the product of $|A|$ and $|\beta|$ is made higher than unity, and when the amplitude reaches a constant value $|V_{out,1}|$, this product becomes equal to unity.

From relations (28) and (29), it follows that in the proposed oscillators, the feedback network coefficient $|\beta|$ is more than unity, and the gain of the active network $|A|$ is less than unity; this is a unique property because in the Colpitts and Hartley oscillators the opposite is true, i.e., $|A| > 1$ and $|\beta| < 1$. Due to this property, the proposed VCO can operate at frequencies significantly exceeding the unity-gain bandwidth of an OPA with sufficiently large oscillation amplitudes.

7. Simulation and Discussion

The circuit of Figure 2a was simulated with the help of Multisim (ed. 14.1) using SPICE models of a real OPA, varactors, and RF coil to confirm the operation efficiency of the proposed VCO topology. We selected ultra-low noise, high-speed OPA LMH6629MF with -3 dB bandwidth of 900 MHz, slew rate of 1600 V/μs, input voltage noise of 0.69 nV/$\sqrt{\text{Hz}}$, and power supply voltage ± 5 V. We also selected UHF varactors BB215 and a 3.3 nH RF coil. The other component values are as follows: $L_1 = L_2 = 1$ μH, $C = 100$ nF, and $R_{dc} = 1$ kΩ.

Figure 7 shows the simulated VCO schematic with chosen component values.

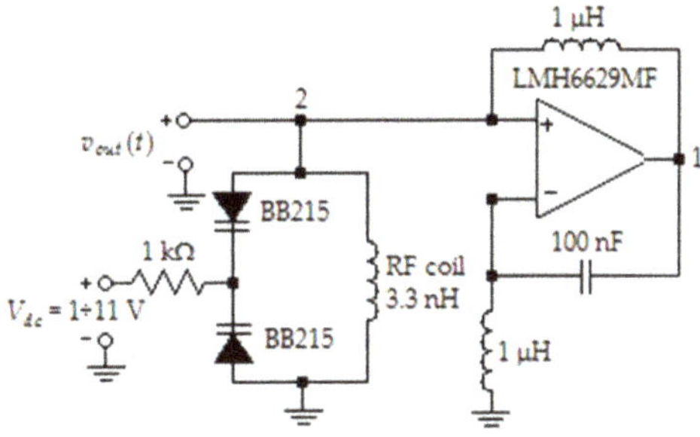

Figure 7. Simulated VCO schematic with indicated component values.

Figure 8 shows the VCO start-up voltage in the interval 0–720 ns at $V_{dc} = 1$ V (a) and $V_{dc} = 11$ V (b). As we can see in Figure 8, the oscillations reach the steady-state condition within 500 ns at $V_{dc} = 1$ V and 50 ns at $V_{dc} = 11$ V. Thus, the starting time of VCO is speedy.

Figure 9 shows the steady-state voltage waveforms at nodes 1 and 2 when $V_{dc} = 1$ V (a) and $V_{dc} = 11$ V (b). As can be seen in Figure 9, the amplitude of the triangular voltage at the output of OPA (node 1) is much less than the amplitude of the sinusoidal voltage at the VCO output (node 2).

From the simulation results shown in Figure 9, it is observed that the THD is 1.5% at $V_{dc} = 1$ V and 1.7% at $V_{dc} = 11$ V, which corresponds to -36.5 dB and -35.4 dB, respectively.

Since in the microwave frequency range, the THD value of -30 dB is considered quite well [23] (p. 291), [24], we can argue that the designed VCO is a low-distortion oscillator.

Let us check Equation (21). From Figure 9a it can be observed that $V_{triangle} = 0.37$ V and $|V_{out,1}| = 3.19$ V. To find the loaded quality factor Q_L, we use the property of the parallel tank circuit that the current circulating in the tank circuit I_{tank} in Q_L times higher than the current in the general circuit I_{res} [21] (p. 911).

Figure 10a illustrates the location of currents I_{tank} and I_{res} in the VCO tank circuit.

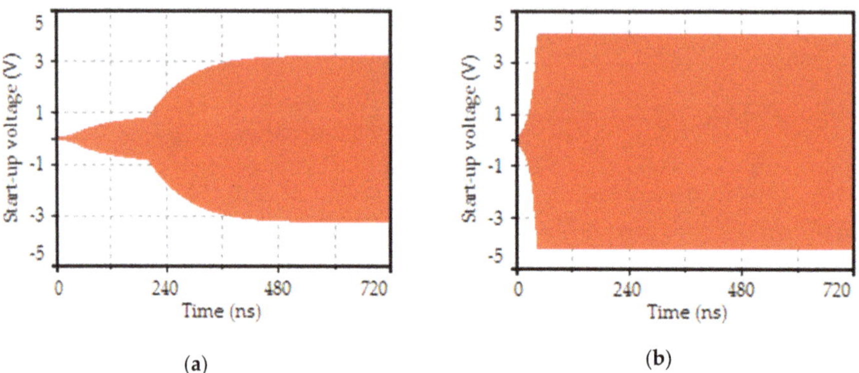

(a)   (b)

Figure 8. VCO start-up voltage (**a**) when $V_{dc} = 1$ V; (**b**) when $V_{dc} = 11$ V.

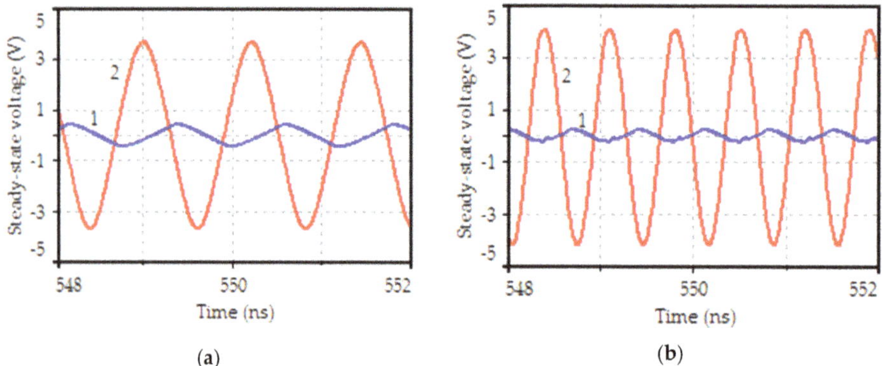

(a)   (b)

Figure 9. Steady-state voltages at nodes 1 and 2; (**a**) when $V_{dc} = 1$ V, (**b**) when $V_{dc} = 11$ V.

Therefore, the loaded quality factor can be calculated by the following equation:

$$Q_L = I_{tank}/I_{res} \tag{30}$$

By simulation, we find that $I_{tank} = 127$ mA and $I_{res} = 15$ mA. Substituting the currents into (30) gives that $Q_L = 8.5$. Further, by substitution of $V_{triangle}$ and Q_L into Equation (21), we calculate that $|V_{out,1}| = 2.76$ V. So, the relative error of amplitude calculation by Equation (21) is only 13.5%.

Figure 10b shows the VCO tuning characteristics. As we can see in Figure 10b, the VCO operates from 830 MHz to 1.429 GHz, i.e., the tunable band is 599 MHz. Thus, the designed circuit is a wideband VCO, which operates in the frequency range, significantly exceeding the bandwidth of the OPA. Indeed, the bandwidth of the LMH6629MF is 900 MHz, but the highest VCO frequency is 1.429 GHz. Using the equation for maximum signal frequency [12], we can find that with the OPA LMH6629MF in the Colpitts or Hartley VCO, the 4 V amplitude of the sinusoidal voltage corresponds to a frequency that is less than 64 MHz.

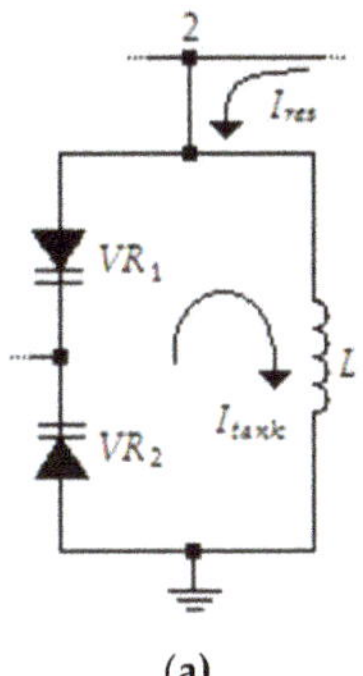

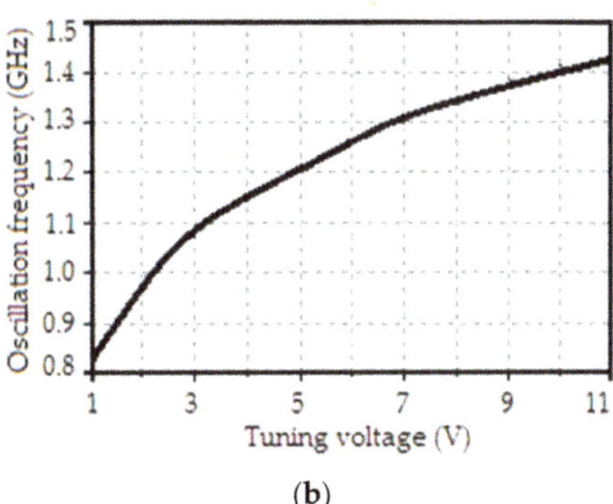

(a)

(b)

Figure 10. (**a**) Illustration of external and contour currents at resonance in the VCO tank circuit; (**b**) VCO tuning characteristics.

Thus, by using the proposed VCO topology, we can increase the maximum operating frequency by more than 22 times. Figure 11 shows the simulated power spectrum of the designed VCO at the lowest frequency (a) and the highest frequency (b). As we can see in Figure 11a, only the second and third harmonics contribute to the THD because the fourth and subsequent harmonics attenuated for at least 90 dB compared to the first harmonic. By analyzing Figure 11b, we find that the second and upper harmonics decrease slowly. However, the second and other harmonics attenuated by at least 36 dB.

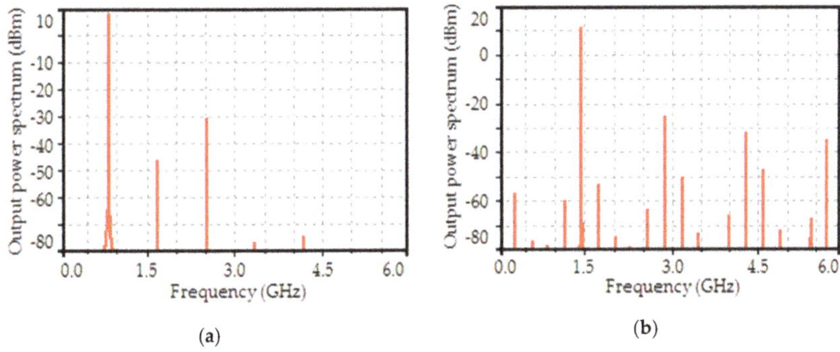

(a)

(b)

Figure 11. Output power spectrum (**a**) when $V_{dc} = 1$ V; (**b**) when $V_{dc} = 11$ V.

One of the essential characteristics of an oscillator operating at microwave frequencies is phase noise. To calculate the phase noise of the simulated VCO, we use the improved Leeson's formula [25] (p. 128)

$$PN(f_m) = 10\log\left\{\frac{FkT}{2P_{out}}\left[\frac{f_{fun}^2 f_c}{f_m^3 4Q_L^2} + \left(\frac{f_{fun}}{2Q_L f_m}\right)^2 + \left(1 + \frac{f_c}{f_m}\right)\right]\right\} \tag{31}$$

where PN is the phase noise (dBc/Hz), F is the noise figure of the oscillator active device (dB), $k \approx 1.38 \times 10^{-23}$ is the Boltzmann constant (J/K), T is the temperature in Kelvin, P_{out} is the oscillator output power, f_{fun} is the frequency of oscillations (Hz), f_c is the $1/f$ corner frequency of active device (Hz), and f_m is the offset frequency (Hz).

According to Reference [7], the noise figure of OPA LMH6629MF is 8 dB. Figure 11a,b shows that $P_{out} = 9$ dBm when $V_{dc} = 1$ V and $P_{out} = 11$ dBm when $V_{dc} = 11$ V. Figure 10b indicates that the minimum frequency of oscillations is 830 MHz, and the maximum is 1429 MHz. From Reference [7] (p.15, Figure 28) we find that f_c = 4 kHz. The loaded quality factor, Q_L, is 8.5 and 2.6 for V_{dc} = 1 V and 11 V, respectively.

Figure 12 shows the dependence of the phase noise versus offset frequency for the VCO shown in Figure 7. As can be seen in Figure 12, the VCO phase noise changes from −153.4 dBc/Hz to −139.3 dBc/Hz at 100 kHz offset frequency when the control voltage varies from 1 V to 11 V. Thus, the maximum in-band phase noise is −139.3 dBc/Hz at f_m = 100 kHz.

Let us compare the characteristics of the simulated VCO with the best VCO designs operating in the microwave frequency range. The commonly used FoM includes phase noise (*PN*), the ratio of f_{fun} to f_m, and the power consumption (P_c) [26].

$$FoM(f_m) = PN(f_m)_{dBc} - 20\log\big(f_{fun}/f_m\big) + 10\log(P_c/1\text{mW}) \tag{32}$$

In Equation (32), the second term allows one to compare the phase noise of the oscillators determined at different frequencies and various frequency offsets. The third term in (32) has a positive sign if P_c > 1 mW and negative if P_c < 1 mW. Thus, the smaller the value of FoM, the higher efficiency of the oscillator.

Table 1 shows a comparison of the designed VCO with microwave oscillators that have been reported in journals and conference proceedings.

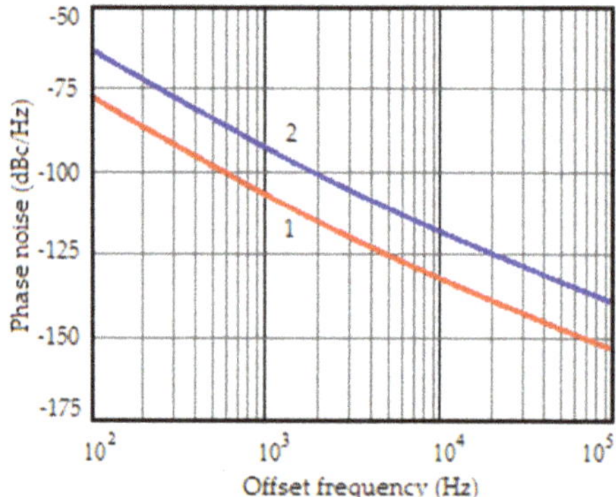

Figure 12. Phase noise versus offset frequency when V_{dc} = 1 V (curve 1) and V_{dc} = 11 V (curve 2).

Table 1. Comparison of designed VCO to other published microwave oscillators.

VCO	Technology	Frequency GHz	Frequency Offset MHz	Phase Noise dBc/Hz	Power Consumption MW	FoM dBc/Hz
[27]	GaN HEMT	4.7	1	−121.7	2.7	−190
[28]	SiGe	29.8	1	−115	37	−185
[29]	HEMT	11.2	1	−97	0.09	−188
[30]	SiGe	6.5 ÷ 15.1	1	−110	19	−177
[31]	GaN HEMT	6.5 ÷ 7.5	1	−110	50	−170
[32]	GaN HEMT	8.82	1	−124.95	21.6	−190.5
[33]	GaN HEMT	7.26	1	−122.48	18.33	−187
[34]	GaN HEMT	1.95	1	−149	400	−189
[35]	GaN HEMT	5.2	1	−125.7	16	−188
[36]	GaAs pHEMT	37.608	1	−112.31	130	−182.7
[37]	SiGe BJT	2.4	1	−128	41	−179.5
[38]	CMOS	26.5	1	−105.8	10.8	−183.9
[39]	CMOS	1.61	0.1	−121	2.7	−202
[40]	CMOS	11.58	1	−112.62	6	−198.6
[41]	CMOS	8	1	−134.3	6.6	−204
This work	SiGe (LMH6629MF)	1.429	0.1	−139.3	240	−198.6

We can see in Table 1 that the designed VCO has the best FoM among oscillators fabricated in GaN HEMT, SiGe, HEMT, GaAs pHEMT, and SiGe BJT technologies where, in general, power consumption is high enough. Moreover, as shown in Table 1, the designed VCO can even compete with the latest achievements in CMOS VCOs; this is due to low phase noise and in spite of higher power dissipation.

8. Experimental Results

We fabricated a prototype of oscillator circuit shown in Figure 2b. Figure 13 shows a PCB assembly of the oscillator. We selected OPA LMH6624 (Texas Instruments) with a slew rate of 350 V/μs, gain bandwidth of 1.5 GHz, supply voltage ±5 V, and the following passive components: $L_0 = 200$ nH (2 × ELJRFR10), $L = 8.2$ nH (high-quality coil ELJQF8N2), and $C_1 = 2.2$ pF (ceramic capacitor C0603C0G1E2R2C030BA). We removed varactors VR_1 and VR_2, and inductor L_2 from the oscillator circuit for simplicity, i.e., $C_{VR1} = C_{VR2} = 0$, and $L_2 = \infty$.

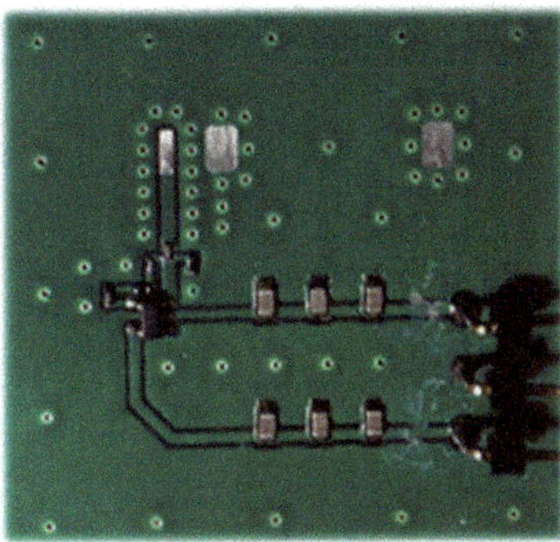

Figure 13. Printed circuit board assembly of fabricated oscillator.

A spectrum analyzer USB-SA44B (Keysight Technologies) was used to measure the output power spectrum. We used RF probes P-20A (Auburn Technology) to connect the oscillator output through a capacitive divider to the spectrum analyzer. Figure 14 shows the block diagram of the measurement experiment.

Figure 15 shows the connection between the core elements of the measurement block diagram. The 2-channel power supply HMC8042-G (Rohde and Schwarz) is not shown in Figure 15.

Figure 16 shows the measured power spectrum. As we can see in Figure 16, the frequency of oscillations is 583.1 MHz. The measured power is −43.9 dBm. However, in estimating the real output power, we should consider 21 dB probes attenuation, 10 dB "Attenuation" setting in the spectrum analyzer, and 13 dB insertion loss of the capacitive divider. Thus, the actual power at the output of the fabricated oscillator was around 0 dBm.

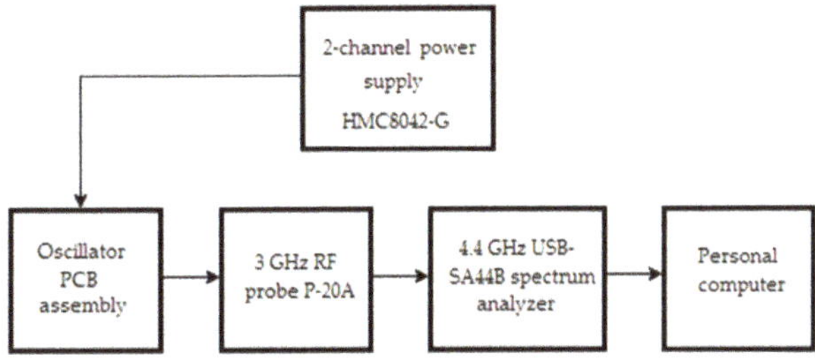

Figure 14. Block diagram of the measurement experiment.

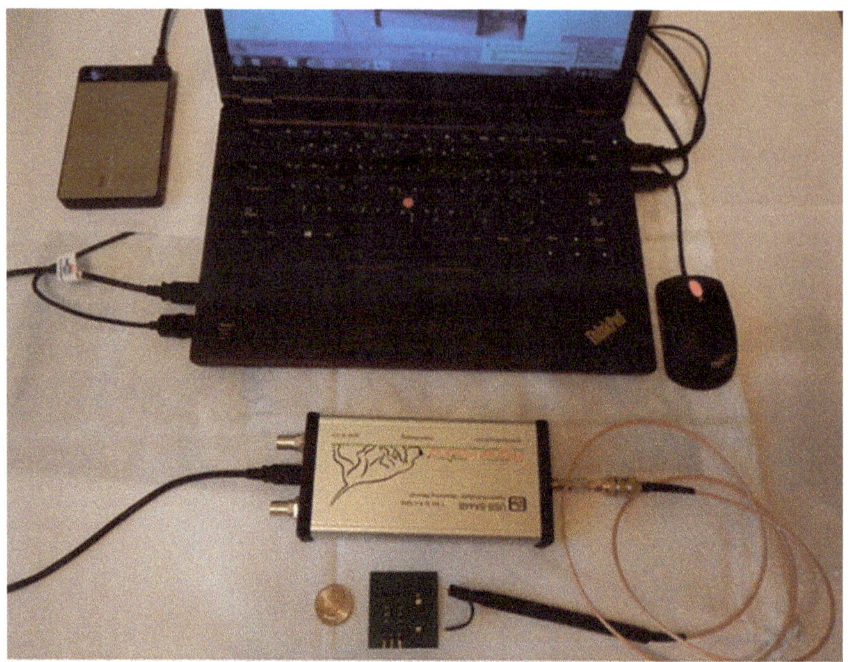

Figure 15. Photograph of the test set-up except for the 2-channel power supply.

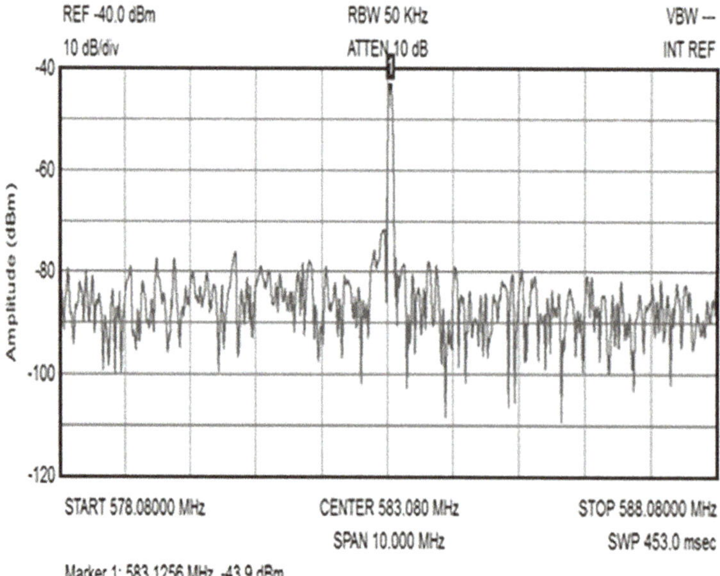

Figure 16. Measured output power spectrum over 10 MHz span and RBW = 50 kHz.

9. Conclusions

This article has proposed a new type of VCO based on resistorless negative impedance converter circuits. A distinctive feature of each proposed VCO circuit is that its output is located at the noninverting input of the OPA; this allows us to significantly increase the frequency and amplitude of the generated oscillations. Mathematical modeling of the amplitude of the oscillations and the start-up conditions were carried out. We have shown that, unlike existing oscillators of the Colpitts and Hartley topology, in the proposed VCO, the gain is less than unity, and the feedback coefficient is greater than unity. Therefore, the proposed oscillators can operate at frequencies exceeding the unity-gain bandwidth. The advantage of the proposed VCO circuit is also the fact that the amplitude of the generated sinusoidal oscillations can be dozens of times higher than the magnitude of the nonsinusoidal oscillations at the output of the OPA. The designed and simulated VCO uses ±5 V supply voltage and operates in a wide frequency range from 830 MHz to 1429 MHz showing a maximum in-band THD of 1.7% while the OPA bandwidth is only 900 MHz; the amplitude of oscillations varies from 3.2 V to 4 V. It has a maximum in-band phase noise of −139.3 dBc/Hz at 100 kHz offset frequency and has an outstanding value of FoM of −198.6 dBc/Hz. The conducted comparison of the designed VCO with VCOs in previously published studies concerning standard FoM has shown that the proposed oscillator is about 8–15 dB more efficient than the GaN HEMT, GaN pHEMT, HEMT, and SiGe VCOs and close to the best CMOS VCOs. We have shown that for the same OPA, the operating frequency of the proposed VCO is 22 times higher than that of the Colpitts and Hartley oscillators. The fabricated prototype-oscillator based on OPA LMH6624 operates at a frequency of 583.1 MHz with a power level of 0 dBm.

The proposed OPA VCO circuits are easy to design. They do not require settings during the manufacturing process. These oscillators can be used from high-frequency to the microwave frequency range. Since the power consumption in the proposed VCO is higher than in the CMOS oscillators, the field of application extends to communications, navigation, and radar equipment with an uninterruptible power supply line. As the proposed schemes do not require tuning, they can be useful in conducting laboratory works in the departments of electronics at universities.

Our future work will include mathematical modeling of phase noise in the proposed oscillators using the method of impulse sensitivity functions.

Author Contributions: This article presents the collective work of two authors. The authors (A.R. and V.U.) jointly participated in the conceptualization of the problem, development of oscillator circuits and mathematical modeling, numerical calculations, and writing the article. V.U. participated in prototype oscillator design and fabrication. A.R. conducted project administration. All authors have read and agreed to the published version of the manuscript.

Funding: This research received no external funding.

Acknowledgments: The authors express thanks to engineer A. Kolesnik for technical support.

Conflicts of Interest: The authors declare no conflict of interest.

Nomenclature

Z_{in}	Input impedance seen by the noninverting terminal of OPA
Z_0	Impedance between the inverting input and output of OPA
Z_1	Impedance between the noninverting input and output of OPA
Z_2	Impedance between the inverting input of OPA and ground
L	Inductor of the tank circuit
VR_1 and VR_2	Varactors of the tank circuit
R_{dc}	Resistor isolating the dc control voltage line from the VCO tank
L_1	Positive feedback inductor
C_1	Positive feedback capacitor
L_0	Negative feedback inductor

C_0	Negative feedback capacitor
L_2	Inductor between inverting terminal of OPA and ground
C_{VCO}	VCO total capacitance
L_{VCO}	VCO total inductance
C_{OPA}	Input capacitance of OPA
C_{PCB}	Parasitic capacitance of PCB
C_{VR1}	Capacitance of varactor VR_1
C_{VR2}	Capacitance of varactor VR_2
y_{11}	Input admittance of feedback network
$i_{OPA}(t)$	Feedback current flowing through impedance Z_1
$v_{out}(t)$	Output voltage of VCO
$I_{OPA,1}$	Complex amplitude of the first harmonic of current $i_{OPA}(t)$
ω_{fun}	Angular frequency of the first (fundamental) harmonic of current $i_{OPA}(t)$
φ_1	Phase of the first harmonic of current $i_{OPA}(t)$
R_{par}	Equivalent parallel resistance of the tank circuit at the fundamental frequency
V_{sat}	Saturation voltage of OPA
R_C	Load resistor connected between OPA output and tank circuit in the Colpitts and Hartley oscillators
$v_{OPA}(t)$	Voltage at the output of OPA (node 1)
V_μ	Complex amplitude of harmonic number μ of voltage $v_{OPA}(t)$
$|V_\mu|$	Amplitude of harmonic number μ of voltage $v_{OPA}(t)$
θ_μ	Phase of harmonic number μ of voltage $v_{OPA}(t)$
$V_{triangle}$	Amplitude of triangular voltage at OPA output
$I_{OPA,\mu}$	Complex amplitude of current harmonic number μ
$|I_{OPA,\mu}|$	Amplitude of current harmonic number μ
φ_μ	Phase of current harmonic number μ
ρ	Tank circuit characteristic impedance
r_s	Series loss resistance of the tank circuit
Q	Quality factor of the parallel tank circuit
Q_L	Quality factor of the loaded parallel tank circuit
$|A|$	Gain of impedance converter
$|\beta|$	Gain of feedback network
φ_A	Initial phase of impedance converter
φ_β	Initial phase of feedback network
$|V_1|$	Voltage amplitude of the first harmonic of triangular voltage at the output of OPA (node 1)
I_{tank}	Current circulating in the tank circuit at resonance
I_{res}	External current entering the tank circuit at resonance
FoM	Figure of merit
P_c	Power consumption of oscillator
PN	Phase noise
P_{out}	Oscillator output power
F	Noise figure
k	Boltzmann constant
T	Temperature in Kelvin
f_c	1/f corner frequency
f_m	Offset frequency
f_{fun}	Frequency of oscillations in Hz

Abbreviations

The following abbreviations exist in the manuscript:

AC	Alternating current
BJT	Bipolar junction transistor
CMOS	Complementary metal-oxide-semiconductor
FoM	Figure of merit
GaAs	Gallium arsenide
GaN	Gallium nitride
HBT	Heterojunction bipolar transistor
HEMT	High-electron-mobility transistor
OPA	Operational amplifier
PCB	Printed circuit board
pHEMT	Pseudomorphic high-electron-mobility transistor
RF	Radio frequency
SiGe	Silicon germanium
SPICE	Simulation program with integrated circuit emphasis
THD	Total harmonic distortion
VCO	Voltage controlled oscillator

References

1. Schulz:, M.; Strobel, A.; Ellinger, F. System considerations and VCO design for a local positioning system at 2.4 GHz for rescue of people on ships and in sea. In Proceedings of the IEEE 10th Workshop on Positioning, Navigation and Communication (WPNC), Dresden, Germany, 20–21 March 2013. [CrossRef]
2. Carlowitz, C.; Esswein, A.; Weigel, R.; Vossiek, M. A low power pulse frequency modulated UWB radar transmitter concept based on switched injection-locked harmonic sampling. In Proceedings of the 7th German Microwave Conference, Ilmenau, Germany, 12–14 March 2012.
3. Scherbina, K.A.; Pechenin, V.V.; Vonsovitch, M.A. Phase-locked loop combined system of voltage controlled oscillator. In Proceedings of the IEEE 9th International Kharkiv Symposium on Physics and Engineering of Microwaves, Millimeter and Submillimeter Waves (MSMW), Kharkiv, Ukraine, 20–24 June 2016. [CrossRef]
4. Grebennikov, A. *RF and Microwave Transistor Oscillator Design*; John Wiley & Sons, Ltd.: Chichester, UK, 2007. [CrossRef]
5. Gonzalez, G. *Foundations of Oscillator Circuit Design*; Artech House: Boston, MA, USA, 2007.
6. Everard, J. *Fundamentals of RF Circuit Design: With Low Noise Oscillators*; John Wiley& Sons, Ltd.: Chichester, UK, 2001. [CrossRef]
7. LMH6629 Ultra-Low Noise, High-Speed Operational Amplifier with Shutdown Datasheet (Rev. I). Available online: http://www.ti.com/lit/ds/symlink/lmh6629.pdf (accessed on 1 December 2014).
8. Ulansky, V.; Machalin, I.; Tkalich, O. Analysis and design of voltage-controlled oscillators using high-speed operational amplifiers. *Proc. Natl. Aviat. Univ.* **2002**, *3*, 171–178. [CrossRef]
9. Ulansky, V.V.; Kolesnik, A.A.; Elsherif, H.M. A new high-performance OPA based VCO for microwave applications. In Proceedings of the 2014 IEEE Microwaves, Radar and Remote Sensing Symposium, Kiev, Ukraine, 23–25 September 2014. [CrossRef]
10. Rashid, M.H. *Microelectronic Circuits. Analysis and Design*; Cengage Learning: Stamford, CT, USA, 2011; p. 883.
11. Jacas, M.M.; Llopis, F. LC sine-wave oscillators using general-purpose voltage operational-amplifiers. *Int. J. Electr. Eng. Educ.* **2007**, *44*, 244–248. [CrossRef]
12. Boylestad, R.; Nashelsky, L. *Electronic Devices and Circuit Theory*, 11th ed.; Pearson: Boston, MA, USA, 2011; p. 619.
13. Ulansky, V.V.; Elsherif, H.M. A voltage-controlled oscillator based on negative inductance converter. In Proceedings of the IEEE 35th International Conference on Electronics and Nanotechnology (ELNANO), Kyiv, Ukraine, 21–24 April 2015. [CrossRef]
14. Chen, W.K. *The Circuits and Filters Handbook*; CRC Press: New York, NY, USA, 2003; pp. 396–397.
15. Horowitz, P.; Hill, W. *The Art of Electronics*, 2nd ed.; Cambridge University Press: Cambridge, UK, 1989; pp. 266–267. [CrossRef]
16. Beck, B.S.; Cunefare, K.A.; Collet, M. The power output and efficiency of a negative capacitance shunt for vibration control of a flexural system. *Smart Mater. Struct.* **2013**, *22*, 1–10. [CrossRef]

17. Negative Impedance Converter. Available online: https://en.wikipedia.org/wiki/Negative_impedance_converter (accessed on 13 April 2020).

18. Savant, C.J.; Roden, M.S.; Carpenter, G.L. *Electronic Design. Circuits and Systems*, 2nd ed.; The Benjamin/Cummings Publishing Company, Inc.: Redwood City, CA, USA, 1991.

19. Johnson, J.B. Thermal agitation of electricity in conductors. *Phys. Rev.* **1928**, *32*, 97. [CrossRef]

20. Nilson, J.W.; Riedel, S.A. *Electric Circuits*, 5th ed.; Addison Wesley: New York, NY, USA, 1996; pp. 791, 806–809.

21. Boylestad, R.L. *Introductory Circuit Analysis*, 10th ed.; Pearson Education: London, UK, 2002.

22. Jaeger, R.C.; Blalock, T.N. *Microelectronic Circuit Design*, 4th ed.; McGraw-Hill Education: New York, NY, USA, 2010; p. 1449.

23. Rhea, R.W. *Oscillator Design and Computer Simulation*; Noble Publishing Corp.: Atlanta, GA, USA, 1996. [CrossRef]

24. Advanced XTAL Products. Available online: https://www.axtal.com/cms/docs/doc101437.pdf (accessed on 19 August 2016).

25. Rohde, U.L.; Poddar, A.K.; Bock, G. *The Design of Modern Microwave Oscillators for Wireless Applications: Theory and Optimization*; John Wiley & Sons, Inc.: New Jersey, NJ, USA, 2005. [CrossRef]

26. Hegazi, E.; Sjoland, H.; Abidi, A. A filtering technique to lower LC oscillator phase noise. *IEEE J. Sol.-St. Circ.* **2001**, *36*, 1921–1930. [CrossRef]

27. Jang, S.-L.; Su, Y.-J.; Lin, K.J.; Wang, B.-J. An 4.7 GHz low power cross-coupled GaN HEMT oscillator. *Microw. Opt. Technol. Lett.* **2018**, *60*, 2442–2447. [CrossRef]

28. Hollmann, A.; Jirovec, D.; Kucharski, M.; Kissinger, D.G.; Fischer, G.; Schreiber, L.R. 30 GHz-voltage controlled oscillator operating at 4 K. *Rev. Sci. Instrum.* **2018**, *89*. [CrossRef] [PubMed]

29. Matheoud, A.V.; Solmaz, N.S.; Boero, G. A low-power microwave HEMT LC oscillator operating down to 1.4 K. *IEEE Trans. Microw. Theory Tech.* **2019**, *67*, 2782–2792. [CrossRef]

30. Drechsel, T.; Joram, N.; Ellinger, F. A 6.5 to 15.1 GHz ultra-wideband SiGe LC VCO with 80% continuous tuning range. In Proceedings of the Eur. Conf. Circuit Theory Design (ECCTD), Catania, Italy, 4–6 September 2017. [CrossRef]

31. Kong, C.; Li, H.; Chen, X.; Jiang, S.; Zhou, J.; Chen, C. A monolithic AlGaN/GaN HEMT VCO using BST thin-film varactor. *IEEE Trans. Microw. Theory Tech.* **2012**, *60*, 3413–3419.

32. Jang, S.L.; Chang, Y.-H.; Chiou, J.-S.; Lai, W.-C. A single GaN HEMT oscillator with four-path inductors. In Proceedings of the 2018 7th International Symposium on Next Generation Electronics (ISNE), Taipei, Taiwan, 7–9 May 2018. [CrossRef]

33. Jang, S.-L.; Chang, Y.-H.; Lai, W.-C. A feedback GaN HEMT oscillator. In Proceedings of the 2018 IEEE International Conference on Microwave and Millimeter Wave Technology (ICMMT), Chengdu, China, 7–11 May 2018. [CrossRef]

34. Lai, S.; Kuylenstierna, D.; Ozen, M.; Horberg, M.; Rorsman, N.; Angelov, I.; Zirath, H. Low phase noise GaN HEMT oscillators with excellent figures of merit. *IEEE Microw. Wirel. Compon. Lett.* **2014**, *24*, 412–414. [CrossRef]

35. Lai, W.-C.; Jang, S.-L.; Chen, Y.-W. Dual-feedback GaN HEMT oscillator. In Proceedings of the 2019 IEEE International Symposium on Radio-Frequency Integration Technology (RFIT), Nanjing, China, 28–30 August 2019. [CrossRef]

36. Chang, C.-L.; Tseng, C.-H.; Chang, H.-Y. A new monolithic Ka-band filter-based voltage-controlled oscillator using 0.15 μm GaAs pHEMT technology. *IEEE Microw. Wirel. Compon. Lett.* **2014**, *24*, 111–113. [CrossRef]

37. Lai, P.W.; Dobos, L.; Long, S. A 2.4 GHz SiGe low phase-noise VCO using on chip tapped inductor. In Proceedings of the 29th European Solid-State Circuits Conference, Estoril, Portugal, 16–18 September 2003. [CrossRef]

38. Fu, Y.; Li, L.; Wang, D.; Wang, X.; He, L. 28-GHz CMOS VCO with capacitive splitting and transformer feedback techniques for 5G communication. *IEEE Trans. Very Large Scale Integr. (VLSI) Syst.* **2019**, *27*, 2088–2095. [CrossRef]

39. Cai, H.L.; Yang, Y.; Qi, N.; Chen, X.; Tian, H.; Song, Z.; Xu, Y.; Zhou, C.; Zhan, J.; Wang, A.; et al. A 2.7-mW 1.36–1.86-GHz LC-VCO with a FOM of 202 dBc/Hz enabled by a 26%-size-reduced nano-particle-magnetic-enhanced inductor. *IEEE Trans. Microw. Theory Tech.* **2014**, *62*, 1221–1228. [CrossRef]

40. Kim, S.J.; Seo, D.I.; Kim, J.S.; Song, R.; Kim, B.S. Compact CMOS LiT VCO achieving 198.6 dBc/Hz FoM. *Electron. Lett.* **2018**, *54*, 175–177. [CrossRef]
41. Zailer, E.; Belostotski, L.; Plume, R. 8-GHz, 6.6-mW LC-VCO with small die area and FOM of 204 dBc/Hz at 1-MHz offset. *IEEE Microw. Wirel. Compon. Lett.* **2016**, *26*, 936–938. [CrossRef]

 *electronics*

Article

Open-Loop Switched-Capacitor Integrator for Low Voltage Applications

Stefano D'Amico [1,*]**, Stefano Marinaci** [1]**, Peter Pridnig** [2] **and Marco Bresciani** [2]

[1] Department of innovation Engineering, University of Salento, 73100 Lecce, Italy;
    stefano.marinaci@unisalento.it
[2] INTEL Austria GmbH, 8, 9524 Villach, Austria; peter.pridnig@intel.com (P.P.);
    marco.bresciani@intel.com (M.B.)
*   Correspondence: stefano.damico@unisalento.it

Received: 13 March 2020; Accepted: 28 April 2020; Published: 6 May 2020

Abstract: An architecture of a switched-capacitor integrator that includes a charge buffer operating in an open-loop is hereby proposed. As for the switched-capacitor filters, the gain of the proposed integrator, which is given by the input/output capacitor ratio, ensures desensitization to process, voltage, and temperature variations. The proposed circuit is suitable for low voltage supplies. It enables a significant power saving compared to a traditional switched-capacitor integrator. This was demonstrated through an analytical comparison between the proposed integrator and a traditional switched-capacitor integrator. The mathematical results were supported and verified by simulations performed on a circuit prototype designed in 16 nm finFET technology with 0.95 V supply. The proposed switched-capacitor integrator consumes 76 µW, resulting in more than twice the efficiency for the traditional closed-loop switched-capacitor filter as an input voltage equal to 31.25 mV at 7 ns clock period is considered. The comparison of architectures was led among the proposed integrator and the state-of-the-art technology in terms of the figure of merit.

Keywords: switched-capacitor filters; low-voltage; finFET

---

1. Introduction

Switched-capacitor filters are used in a variety of applications like sensors interfaces [1], audio applications [2], RF front-ends [3], analog-to-digital converters [4], etc. The high accuracy, comparable to the capacitors mismatch and the ease of reprogramming, makes this topology preferable. Such filters require high-performance operational transconductance amplifiers (OTAs). Using modern IC technologies helps to reach high-frequency operation. However, obtaining high DC-gain becomes more challenging for two main reasons: the low supply voltage of the modern IC technologies that limit the number of stackable devices, generally used to obtain high DC-gain OTAs; the reduced length of the transistors that reduces the output resistances and limits the transistor intrinsic gain. On the other hand, the required high-DC gain is obtainable, increasing the number of gain stages of the OTA at the cost of higher power consumption.

In this paper, an open-loop switched-capacitor filter as a building block for the design of a higher-order switched-capacitor filter is presented. Despite its open-loop architecture, the integrator gain depends on the capacitor ratio whose accuracy is related to the capacitor mismatch as it is usually done in closed-loop switched-capacitor integrators. Furthermore, the proposed switched-capacitor integrator enables low voltage operation and relaxes the power requirement compared to the closed-loop counterpart. The paper is organized as follows: Section 2 starts with the analysis from the conventional switched-capacitor integrator used as a benchmark. Section 3 extends the analysis to the proposed open-loop integrator. Section 4 reports the simulation results and Section 5 concludes the paper.

2. The Closed Loop Switched-Capacitor Integrator

Figure 1 shows the conventional architecture of a switched capacitor integrator.

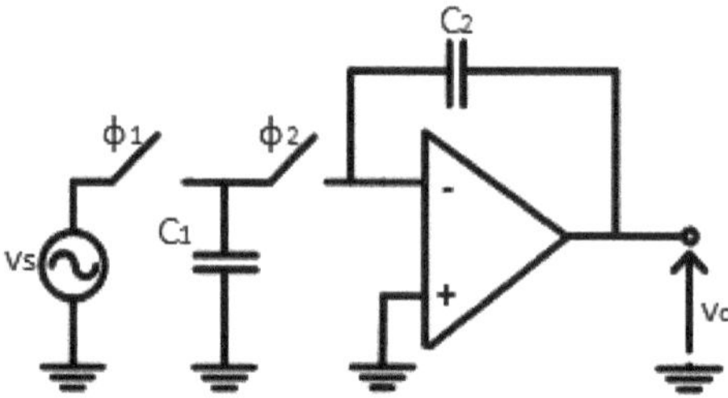

Figure 1. Conventional architecture of a closed-loop switched-capacitor integrator.

The switching scheme of this architecture is defined by two complementary clock phases, ϕ_1 and ϕ_2. During ϕ_1 phase, the input signal, v_s, is sampled with the input capacitor, C_1. During ϕ_2 phase the charge collected by C_1 is transferred to the feedback capacitor C_2, assuming an ideal virtual ground at the input of the OTA. The overall transfer function in Z-domain is the following

$$\frac{v_o}{v_s}(z) = -\frac{C_1}{C_2} \cdot \frac{z^{-1}}{1 - z^{-1}} \tag{1}$$

As in many other electronic systems, the feedback in this circuit serves two main functions:

- mitigate the impact of nonlinearities in the OTA;
- desensitize the overall transfer function to process, voltage, and temperature (PVT) variations.

The cost of these desirable features is an excessive OTA requirement.

2.1. Analysis of the Requirements of the Single Stage OTA

To evaluate the OTA requirements, a single-stage architecture was considered. The linear model of the OTA includes a transconductance g_m and an output resistance r_o. Figure 2 reports the linear model of the integrator including the model of the OTA.

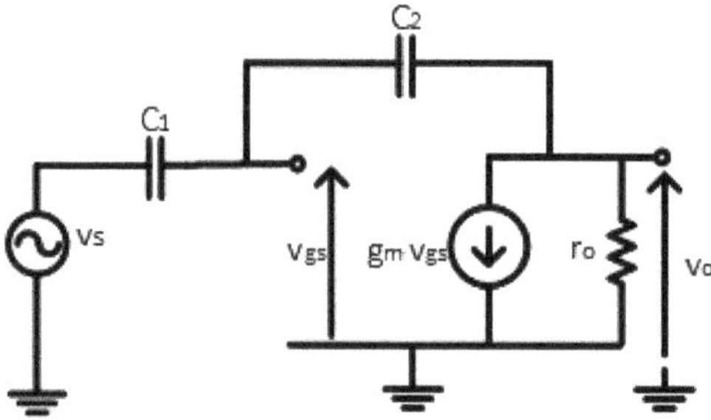

Figure 2. Equivalent small-signal circuit of the closed-loop switched-capacitor integrator with a single-stage operational transconductance amplifier (OTA).

In a switched-capacitor integrator, the input signal, v_s, changes suddenly at each clock hit. Therefore, v_s can be assimilated to a step signal whose a maximum amplitude is equal to V_i:

$$v_s(t) = V_i \cdot u(t) \tag{2}$$

where $u(t)$ is the unitary step signal.

The approach followed for this analysis is the following:

1. firstly, the transfer function $\frac{v_o}{v_s}$ is calculated;

2. the output voltage $v_0(s)$ is calculated in s domain multiplying the transfer function $\frac{v_o}{v_s}$ and the Laplace transform of $v_s(t)$;

3. $v_0(s)$ is then inverse transformed to get the output voltage, $v_0(t)$, in time domain.

The transfer function, $\frac{v_o}{v_s}$, is calculated as follows:

$$\frac{v_o}{v_s} = -\frac{C_1}{C_2 \cdot \left(1 + \frac{1 + \frac{C_1}{C_2}}{g_m \cdot r_o}\right)} \cdot \frac{1 - \frac{s \cdot C_2}{g_m}}{1 + \frac{s \cdot C_1}{g_m \cdot \left(1 + \frac{1 + \frac{C_1}{C_2}}{g_m \cdot r_o}\right)}} = -\frac{C_1}{C_2 \cdot \left(1 + \frac{1 + \frac{C_1}{C_2}}{A_o}\right)} \cdot \frac{1 - \frac{s \cdot C_2}{g_m}}{1 + \frac{s \cdot C_1}{g_m \cdot \left(1 + \frac{1 + \frac{C_1}{C_2}}{A_o}\right)}} \tag{3}$$

where A_0 ($= g_m \cdot r_o$) is the OTA DC-gain. Assuming that both r_o and g_m tend to infinite, $\frac{v_o}{v_s}$ can be approximate to the ideal value $\left.\frac{v_o}{v_s}\right|_{ideal}$:

$$\left.\frac{v_o}{v_s}\right|_{ideal} = \left.\frac{v_o}{v_s}\right|_{\substack{r_o \to \infty \\ g_m \to \infty}} = -\frac{C_1}{C_2} \tag{4}$$

The function $v_s(t)$ reported in Equation (2) is transformed in s domain and combined with Equation (3), and then the inverse Laplace Transform is evaluated as follows:

$$v_0(t) = -\frac{C_1}{C_2 \left(1 + \frac{1 + \frac{C_1}{C_2}}{A_o}\right)} \cdot V_i \cdot \left(1 - \frac{\tau_z + \tau_p}{\tau_p} \cdot e^{-\frac{t}{\tau_p}}\right) \tag{5}$$

where τ_p and τ_z are time constants calculated as reciprocal of pole and zero of the transfer function, i.e.,

$$\tau_p = \frac{C_1}{g_m \cdot \left(1 + \frac{1 + \frac{C_1}{C_2}}{A_o}\right)} \quad \text{and} \quad \tau_z = \frac{C_2}{g_m} \tag{6}$$

Assuming $A_0 \gg 1 + \frac{C_1}{C_2}$, τ_p can be approximated as follows:

$$\tau_p \cong \frac{C_1}{g_m} \tag{7}$$

The output voltage $v_0(0)$ at $t = 0$, is defined as:

$$v_0(0) = -\frac{C_1}{C_2 \cdot \left(1 + \frac{1 + \frac{C_1}{C_2}}{A_o}\right)} \cdot V_i \cdot \left(1 - \frac{\tau_z + \tau_p}{\tau_p}\right) \cong V_i \tag{8}$$

The discontinuity is due to the zero in the transfer function.

2.2. Requirements of the Single Stage OTA

A finite gain of the OTA, A_0, and the non-null time is required for settling introduced errors on the output voltage. The OTA finite gain determines an error in a steady state. This error is called static error, ε_{stat}:

$$\varepsilon_{stat} = \left| \frac{v_o}{v_s} \right|_{ideal} V_i - v_o(t \to \infty) \right| \tag{9}$$

On the base of Equations (4) and (5), the static error, ε_{stat}, is calculated as follows:

$$\varepsilon_{stat} = \left| \left(\frac{C_1}{C_2} - \frac{C_1}{C_2\left(1 + \frac{1+\frac{C_1}{C_2}}{A_o}\right)} \right) \cdot V_i \right| = \frac{\frac{C_1}{C_2} \cdot \frac{1+\frac{C_1}{C_2}}{A_o}}{1 + \frac{1+\frac{C_1}{C_2}}{A_o}} \cdot V_i \cong \frac{C_1}{C_2} \cdot \frac{1 + \frac{C_1}{C_2}}{A_o} \cdot V_i \tag{10}$$

where the last approximation is valid as $A_0 \gg 1 + \frac{C_1}{C_2}$.

The charging process of the feedback capacitance, C_1, has a finite duration. In the following calculations, it is assumed that the duration of the charging phase is half of the clock period, T_{CLK}. Furthermore, an incomplete settling of the output voltage produces an error, which is called dynamic error, ε_{dyn}, which is defined as follows:

$$\varepsilon_{dyn} = \left| v_o(t \to \infty) - v_o\left(\frac{T_{CLK}}{2}\right) \right| \tag{11}$$

By using the expression of $v_o(t)$, reported in Equation (5), the dynamic error, ε_{dyn}, is calculated as follows:

$$\varepsilon_{dyn} = \frac{C_1}{C_2} \cdot V_i \cdot \frac{1}{1 + \frac{1+\frac{C_1}{C_2}}{A_o}} \cdot \frac{\tau_z + \tau_p}{\tau_p} \cdot e^{-\frac{T_{CLK}}{2\tau_p}} \tag{12}$$

which can be simplified combining Equation (12) with Equations (6) and (7) that:

$$\varepsilon_{dyn} = \left(1 + \frac{C_1}{C_2}\right) \cdot V_i \cdot \frac{1}{1 + \frac{1+\frac{C_1}{C_2}}{A_o}} e^{-\frac{T_{CLK}}{2\tau_p}} \tag{13}$$

Figure 3 shows the qualitative behavior of the output voltage. Both static and dynamic errors are highlighted.

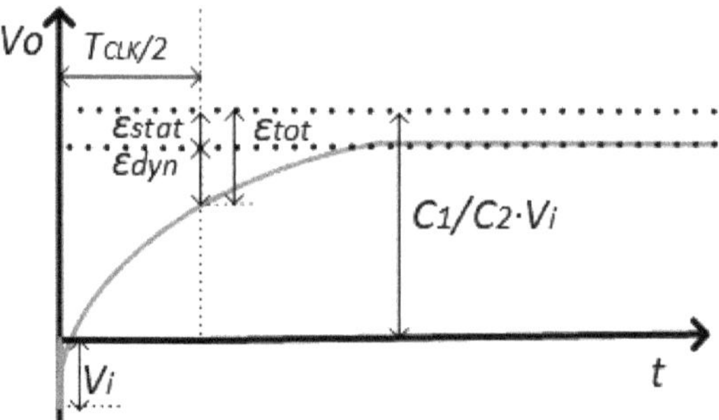

Figure 3. Output voltage behavior of the closed-loop switched-capacitor integrator in the linear regime.

The overall error, ε_{tot}, evalueated on the output voltage, is defined as the difference between the ideal output voltage, $C_2/C_1 \cdot V_i$, and the output voltage measured at $T_{CLK}/2$:

$$\varepsilon_{tot} = \left| \frac{C_2}{C_1} \cdot V_i - v_0\left(\frac{T_{CLK}}{2}\right) \right| = \varepsilon_{stat} + \varepsilon_{dyn} \tag{14}$$

As Equation (14) shows, ε_{tot} is calculated as the sum of ε_{stat} and ε_{dyn}.

2.2.1. DC-Gain Requirement of the Single-Stage OTA

The overall error must be less than the required accuracy, ξ, which is a parameter related to the application. According to the definition, both ε_{stat} and ε_{dyn} must be positive and smaller than the required accuracy, ξ.

To fulfill the DC-gain requirement, the accuracy of the static error, ε_{stat}, needs to be addressed as follows:

$$\varepsilon_{stat} < \xi \tag{15}$$

Combining Equations (10) and (15), the following is obtained:

$$\frac{C_1}{C_2} \cdot \frac{1 + \frac{C_1}{C_2}}{A_o} \cdot V_i < \xi \tag{16}$$

Then, the following constraint on the DC-gain, A_o, is derived:

$$A_o > \frac{C_1^2}{C_2^2} \cdot \frac{V_i}{\xi} - 1 \cong \frac{C_1^2}{C_2^2} \cdot \frac{V_i}{\xi} \tag{17}$$

Since the DC-gain, A_o, undergoes process, voltage and temperature variations, the sensitivity of A_o, $S_{A_0}^{v_0\left(\frac{T_{CLK}}{2}\right)}$, of the output voltage at $t = \frac{T_{CLK}}{2}$, $v_0\left(\frac{T_{CLK}}{2}\right)$, is evaluated from Equation (5) as follows:

$$S_{A_0}^{v_0\left(\frac{T_{CLK}}{2}\right)} = \frac{A_0}{v_0\left(\frac{T_{CLK}}{2}\right)} \cdot \frac{\partial v_0\left(\frac{T_{CLK}}{2}\right)}{\partial A_0} = A_0 \cdot \frac{1 + \frac{C_1}{C_2}}{\left(A_0 + \frac{C_1}{C_2} + 1\right)^2} \cong \frac{1 + \frac{C_1}{C_2}}{A_0} \tag{18}$$

where last approximation is valid as $A_0 \gg 1 + \frac{C_1}{C_2}$.

2.2.2. Transconductance Requirement of the Single Stage OTA

A transconductance constrain is determined through the relation between the accuracy specification and the dynamic error, ε_{dyn}, i.e.:

$$\varepsilon_{dyn} < \xi \tag{19}$$

Combining Equations (13) and (19), it is obtained:

$$\left(1 + \frac{C_1}{C_2}\right) \cdot V_i \frac{1}{1 + \frac{1 + \frac{C_1}{C_2}}{A_o}} \cdot e^{-\frac{T_{CLK}}{2 \cdot \tau_p}} \cong \left(1 + \frac{C_1}{C_2}\right) \cdot V_i \cdot e^{-\frac{T_{CLK}}{2 \cdot \tau_p}} < \xi \tag{20}$$

The approximation contained in Equation (20) is justified as it is assumed that $A_0 \gg 1 + \frac{C_1}{C_2}$. Combining Equations (7) and (20) the following constraint on g_m is set:

$$g_m > \frac{2 \cdot C_1}{T_{CLK}} \cdot \ln\left(\frac{V_i}{\xi} \cdot \left(1 + \frac{C_1}{C_2}\right)\right) \tag{21}$$

As for the DC-gain, A_o, the sensitivity of the output voltage, $S_{g_m}^{v_o\left(\frac{T_{CLK}}{2}\right)}$, at $\frac{T_{CLK}}{2}$, $v_o\left(\frac{T_{CLK}}{2}\right)$, with respect to g_m is derived from Equation (5) as follows:

$$S_{g_m}^{v_o\left(\frac{T_{CLK}}{2}\right)} = \frac{g_m}{v_o\left(\frac{T_{CLK}}{2}\right)} \cdot \frac{\partial v_o\left(\frac{T_{CLK}}{2}\right)}{\partial g_m} = \frac{T_{CLK}}{2\cdot\tau_p} \cdot \frac{\left(1 + \frac{2\tau_z}{T_{CLK}} + \frac{C_2}{C_1}\right)e^{-\frac{T_{CLK}}{2\tau_p}}}{1 - \left(1 + \frac{C_2}{C_1}\right)e^{-\frac{T_{CLK}}{2\tau_p}}} \cong \frac{T_{CLK}}{2\cdot\tau_p} \cdot \left(1 + \frac{C_2}{C_1}\right) \cdot e^{\frac{-T_{CLK}}{2\tau_p}} \tag{22}$$

where the last approximation is valid as $\left(1 + \frac{C_2}{C_1}\right) \cdot e^{\frac{-T_{CLK}}{2\tau_p}} \ll 1$ and $\tau_z \ll \frac{T_{CLK}}{2}$.

2.3. Circuit Implementation of the Single Stage OTA

A common circuit solution for the OTA is represented by the telescopic Cascode OTA, shown in Figure 4 [5,6]. As the DC-gain requirement is satisfied and the output signal swing is sufficient, the telescopic Cascode OTA reported in Figure 2 remains the most efficient and simplest OTA solution. Therefore, it was used as a benchmark in this paper.

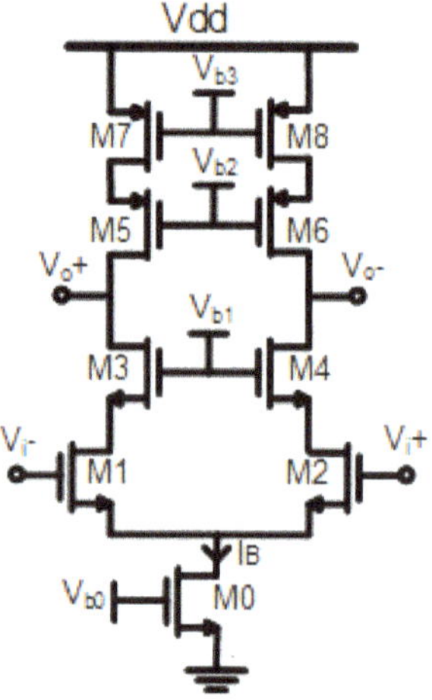

Figure 4. Telescopic Cascode OTA.

The overdrive voltage of M_1-M_2 input transistors is limited by the available supply voltage, V_{dd}, and the NMOS transistor threshold, V_{THN}. Assuming a common mode, V_{cm}, equal to $V_{dd}/2$, by applying Kirchoff's voltage law we obtain:

$$V_{cm} = \frac{V_{dd}}{2} = V_{GS1} + V_{DS0} \tag{23}$$

where V_{GS1} is the gate-source voltage of M_1-M_2 input transistors, and V_{DS0} is the drain-source voltage of the M_0 bias transistor.

The common mode, V_{cm}, must assure that M_1-M_2 and the M_0 transistors work in the saturation region. Therefore, assuming that all the overdrives of M_1-M_2 and M_0 transistors are equal to V_{ov}, from Equation (23) we derive:

$$\frac{V_{dd}}{2} > V_{THN} + 2\cdot V_{ov} \tag{24}$$

Thus:

$$V_{ov} < \frac{\frac{V_{dd}}{2} - V_{THN}}{2} \tag{25}$$

As seen in Equation (25), there is a strict limitation to the design of the overdrive of the input transistors at low supply voltage, which is typical of the modern CMOS IC technologies. For example, in finFET 16 nm technology, V_{dd} is 0.95 V, and V_{THN} is 0.275 V, therefore V_{ov} must be less than 100 mV.

2.4. Small Signal Analysis of the Single-Stage OTA

The transconductance g_m of the linear model of Figure 2 corresponds to the transconductance g_{m1} of M_1-M_2 input transistors. The bias current, I_B, of the OTA is defined by the settling requirements, which mainly depends on M_1-M_2 input transistors.

The output resistance r_o of the linear model of Figure 2 is calculated as follows:

$$r_o = g_{m3} \cdot r_{o3} \cdot r_{o1} \| g_{m5} \cdot r_{o5} \cdot r_{o7} \cong \frac{1}{2} g_{m3} \cdot r_{o3} \cdot r_{o1} \tag{26}$$

where g_{m3} and g_{m5} are the transconductances of M_3 and M_5 transistors, respectively, and r_{o1}, r_{o3}, and r_{o5} are the output resistances of M_1, M_2, and M_3 transistors, respectively. The last approximation in Equation (24) is valid assuming $g_{m3} \cong g_{m5}$, $r_{o3} \cong r_{o5}$, and $r_{o1} \cong r_{o7}$. In practice, the output resistance of M_1-M_2 transistors, r_{o1}, is boosted by the intrinsic gain of transistor M_3, $g_{m3} \cdot r_{o3}$.

The voltage gain, A_0, can be calculated as follows:

$$A_0 = g_{m1} \cdot r_o \cong \frac{1}{2} g_{m1} \cdot g_{m3} \cdot r_{o3} \cdot r_{o1} \tag{27}$$

Rearranging Equation (27), we obtain:

$$A_0 \cong 2 \cdot \frac{V_{A3} \cdot V_{A1}}{V_{ov3} \cdot V_{ov1}} \tag{28}$$

where V_{A3}, V_{A1}, V_{ov3}, and V_{ov1} are the early and the overdrive voltages of M_3 and M_1 transistors, respectively. As derived in Equation (28), the margins to increase the voltage gain A_o are limited. A possibility to increment the voltage gain consists of reducing V_{ov3} and V_{ov1}. M_3 and M_1 transistors are then pushed to work in the subthreshold region, where the transconductance depends only on the bias current, while the transistor overdrives approach their inferior limit of about 50 mV [7]. Therefore, V_{ov1} and V_{ov3} have a strict range of variability between 50 mV and 100 mV. V_{A3} and V_{A1} can be increased by augmenting the length of M_3 and M_1. This solution degrades the frequency performance of the OTA since larger transistors introduce bigger parasitic capacitances. Moreover, every IC fabrication process has an intrinsic limit to the maximum allowable transistor length, which is lower and lower as the technology is scaled. For example, in FinFET 16 nm, the maximum allowable transistor length is 240 nm.

If the DC-gain requirement is not reachable by the telescopic Cascode OTA, it is necessary to modify the OTA architecture. An increment of r_o and, consequently, A_o is obtained by using a regulated Cascode OTA [8] or adding an output stage [9]. Both previous solutions imply a significant increase in power consumption. It is also possible to increase the gain by augmenting the number of stacked transistors. At low voltage supply, the last solution is not practicable because of the further reduction of the output signal swing.

2.5. Slew-Rate (SR) Analysis of the Single-Stage OTA

Due to the finite bias current, I_B, the OTA goes into a slew-rate regime at the beginning of the charging process.

According to Equation (5), the maximum rate of variation of the output voltage is obtained at $t = 0$ s.

$$\left| \frac{dv_o(t)}{dt} \right|_{max} = \frac{C_1}{C_2 \cdot \left(1 + \frac{1 + \frac{C_1}{C_2}}{A_o} \right)} \cdot V_{SRi,max} \cdot \frac{\tau_z + \tau_p}{\tau_p^2} = \frac{V_{SRo,max}}{\tau_p} \tag{29}$$

where $V_{SRi,max}$ is the maximum amplitude of the input voltage step that keeps the OTA in the linear region. The corresponding output voltage in steady state is given by $V_{SRo,max}$, which is calculated as follows:

$$V_{SRo,max} = \frac{C_1}{C_2 \cdot \left(1 + \frac{1 + \frac{C_1}{C_2}}{A_o}\right)} \cdot \left(1 + \frac{C_2}{C_1}\right) \cdot V_{SRi,max} \cong \left(1 + \frac{C_1}{C_2}\right) \cdot V_{SRi,max} \tag{30}$$

where the last approximation is valid as $A_0 \gg 1 + C_1/C_2$. In a single-stage OTA, the slew-rate depends on the bias current I_B and the feedback capacitance C_2, i.e.:

$$SR = \frac{I_B}{C_2} = \frac{I_B \cdot g_{m1}}{C_2 \cdot g_{m1}} = \frac{V_{ov1}}{\tau_z} \tag{31}$$

where g_{m1} and V_{ov1} are the transconductance and the overdrive of the M_1-M_2 input transistors, respectively. Matching the SR formula in Equation (31) to the maximum rate of variation of the output voltage reported in Equation (29), the $V_{SRo,max}$ calculation is obtained:

$$V_{SRo,max} = V_{ov1} \cdot \frac{\tau_p}{\tau_z} \cong V_{ov1} \cdot \frac{C_1}{C_2} \tag{32}$$

Due to its differential structure, the OTA starts slewing as the input differential voltage step, V_i, overcomes $2 \cdot V_{SRi,max}$. The value of $V_{SRi,max}$ is calculated by combining Equations (30) and (32):

$$V_{SRi,max} \cong \frac{1}{\left(1 + \frac{C_1}{C_2}\right)} \cdot V_{SRo,max} = \frac{V_{ov1}}{\left(1 + \frac{C_2}{C_1}\right)} \tag{33}$$

The OTA slews until the output voltage reaches the value $\frac{-C_1}{C_2} V_i + 2 V_{SRo,max}$:

$$V_i - \frac{2 \cdot V_{SRo,max}}{\tau_p} t \big|_{t=\tau_s} = -\frac{C_1}{C_2} V_i + 2 \cdot V_{SRo,max} \tag{34}$$

where the starting value of V_i depends on the fact that, at $t = 0$ s, the capacitances of the integrator behave like short circuits, transferring the input voltage directly to the output.

From the previous equation, it is possible to calculate the slewing time of the OTA, τ_s:

$$\tau_s = \tau_p \cdot \left(\left(1 + \frac{C_1}{C_2}\right) \cdot \frac{V_i}{2 \cdot V_{SRo,max}} - 1\right) = \tau_p \cdot \left(\frac{V_i}{2 \cdot V_{SRi,max}} - 1\right) \tag{35}$$

During τ_s, the OTA output voltage evolves according to the linear law. Considering the slew-rate, the equation of the differential output voltage, $v_{od}(t)$, is then calculated as follows:

$$\begin{cases} v_{od}(t) = V_i - \frac{2 \cdot V_{SRo,max}}{\tau_p} \cdot t & \text{for } 0 < t < \tau_s \\[2mm] v_{od}(t) = -\frac{C_1}{C_2 \cdot \left(1 + \frac{1 + \frac{C_1}{C_2}}{A_o}\right)} \cdot V_i + 2 \cdot V_{SRo,max} \cdot e^{-\frac{t - \tau_s}{\tau_p}} \cong -\frac{C_1}{C_2} \cdot V_i + 2 \cdot V_{SRo,max} e^{-\frac{t - \tau_s}{\tau_p}} & \text{for } t > \tau_s \end{cases} \tag{36}$$

Figure 5 shows the step response of the closed loop switched capacitor integrator including the slewing period.

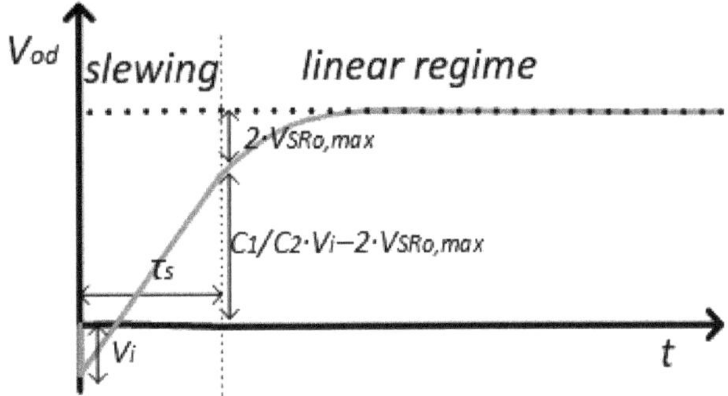

Figure 5. Output voltage behavior of the closed-loop switched-capacitor integrator including the slewing period.

2.6. Signal to Noise (SNR) Calculations of the Closed-Loop Switched-Capacitor Integrator

The telescopic Cascode OTA suffers from a reduced output swing. Indeed, both single-ended output voltages must guarantee that the Cascode transistors (M_3-M_4 and M_5-M_6) work in a saturation region even under the signal swing. The main limitation is the negative output swing since three transistors are stacked between the ground and the output nodes, while only two transistors are stacked between V_{dd} and the output nodes. Focusing the analysis on a single branch, V_{o+} must satisfy the following inequation to guarantee that M_3 transistors operate in saturation region:

$$V_{o+} > V_{DS,sat3} + V_{S3} = V_{ov} + V_{S3} \tag{37}$$

where V_{S3} and $V_{DS,sat3}$ are the source and the saturation voltages of M_3 transistor, respectively. It is assumed that $V_{ds,sat3}$ is equal to V_{ov}. The bias voltage V_{b1} is chosen to make M_1 transistor operating in saturation, i.e.:

$$V_{DS1} = V_{S3} - V_{S1} > V_{DS,sat1} = V_{ov} \tag{38}$$

where V_{DS1}, V_{S1}, and $V_{DS,sat1}$ are the drain-source, the source, and the saturation voltages of M_1 transistor, respectively. In this case, it is assumed that $V_{ds,sat1}$ is equal to V_{ov}. V_{S1} is derived from the input transistor common mode, V_{cm}, by dropping the gate-drain voltage of the M_1 transistors, V_{GS1}, i.e.:

$$V_{S1} = V_{cm} - V_{GS1} = V_{cm} - V_{TH} - V_{ov} \tag{39}$$

Combining Equations (38) and (39) we obtain the minimum source voltage of M_3 transistor, $V_{S3,min}$:

$$V_{S3} > V_{S1} + V_{ov} = V_{S3,min} \tag{40}$$

By replacing V_{S3} in Equation (37) with the value of $V_{S3,min}$ calculated in Equation (40), the minimum value of V_{o+}, $V_{o+,min}$, is obtained:

$$V_{o+} > V_{ov} + V_{S3} = V_{cm} - V_{THN} = V_{o+,min} \tag{41}$$

As the output voltage starts swinging from the common-mode voltage, V_{cm}, down to $V_{o+,min}$, it is possible to calculate the maximum output voltage swing, V_{swing}:

$$V_{swing} = 2 \cdot (V_{cm} - V_{o+,min}) = 2 \cdot V_{THN} \tag{42}$$

where the 2 factor is due to the differential architecture.

The thermal noise due to the switches around C_1 is calculated as $2 \cdot \frac{K \cdot T}{C_1}$, where the coefficient 2 takes into account both the sampling (ϕ_1) and the integration phase (ϕ_2). Assuming that the thermal noise $2 \cdot \frac{K \cdot T}{C_1}$ is dominant, from Equation (42), the signal to noise ratio of the overall closed-loop switched-capacitor integrator, SNR_{CL}, is calculated as follows:

$$SNR_{CL} = \frac{1}{2} \cdot \frac{V_{swing}^2}{\overline{v_{o,n}^2}} = \frac{1}{2} \cdot \frac{4 \cdot V_{THN}^2}{2 \cdot \frac{K \cdot T}{C_1} \cdot \left(\frac{C_1}{C_2}\right)^2} = \frac{V_{THN}^2 \cdot C_2^2}{K \cdot T \cdot C_1} \tag{43}$$

where $\overline{v_{n,o}^2}$ is the total output noise, as a result of the thermal noise contribution due to the C_1 switched-capacitor multiplied by the square of the integrator gain $\left(\frac{C_1}{C_2}\right)^2$, furthermore, the $\frac{1}{2}$ factor takes into account that the input signal is a sinusoid.

2.7. Power Consumption Requirements

Regarding the telescopic Cascode shown in Figure 4, the power consumption is given by the product of the supply voltage, V_{dd}, and the bias current I_B:

$$P_{w,tot} = V_{dd} \cdot I_B \tag{44}$$

Assuming dominant the thermal noise of C_1, the power consumption of the switched-capacitor integrator is determined by the settling time requirement. In fact, as the input transistor overdrive, V_{ov1}, is bonded to considerations on the DC-point at low voltage supply, the constraint on the input transistor transconductance, g_{m1}, expressed by in Equation (21), determines the minimum required bias current $I_{B,min}$:

$$I_{B,min} = \frac{2 \cdot C_1}{T_{CLK}} \cdot V_{ov1} \cdot \ln\left(\frac{V_i}{\xi} \cdot \left(1 + \frac{C_1}{C_2}\right)\right) \tag{45}$$

Therefore, the minimum power consumption, $P_{w,min}$, is obtained as follows:

$$P_{w,min} = V_{dd} \cdot I_{B,min} = V_{dd} \cdot \frac{2 \cdot C_1}{T_{CLK}} \cdot V_{ov1} \cdot \ln\left(\frac{V_i}{\xi} \cdot \left(1 + \frac{C_1}{C_2}\right)\right) \tag{46}$$

As the OTA starts slewing, the minimum bias current, $I_{B,min}$, is determined by taking into account a different calculation for the dynamic error, ε_{dyn}. Indeed, considering Equation (36) that assumes the slewing of the OTA, the differential output voltage at $t = \frac{T_{CLK}}{2}$ is calculated as follows:

$$v_{od}\left(\frac{T_{CLK}}{2}\right) \cong -\frac{C_1}{C_2} \cdot V_i + 2 \cdot V_{SRo,max} \cdot e^{-\frac{\frac{T_{CLK}}{2} - \tau_s}{\tau_p}} \tag{47}$$

The dynamic error, ε_{dyn}, is, then, calculated as follows:

$$\varepsilon_{dyn} = \left| v_{od}(t \to \infty) - v_o\left(\frac{T_{CLK}}{2}\right) \right| = 2 \cdot V_{SRo,max} e^{-\frac{\frac{T_{CLK}}{2} - \tau_s}{\tau_p}} \tag{48}$$

Since ε_{dyn} must be less than the required accuracy, ξ, as reported in Equation (19), we obtain:

$$\tau_p < \frac{\frac{T_{CLK}}{2} - \tau_s}{\ln\left(\frac{2V_{SRo,max}}{\xi}\right)} \tag{49}$$

Moreover, the minimum bias current $I_{B,min}$ is evaluated considering τ_p reported in Equation (7), the transconductance of the input transistors, g_{m1}, determined as $\frac{I_{B,min}}{V_{ov1}}$, the formula of the slewing time, τ_s, in Equation (35), and the previous equation. As a result the minimum bias current $I_{B,min}$, is calculated as follows:

$$I_{B,min} = \frac{2 \cdot V_{ov1} \cdot C_1}{T_{CLK}} \cdot \left(ln\left(\frac{2 \cdot V_{SRo,max}}{\xi}\right) + \frac{V_i}{2 \cdot V_{SRi,max}} - 1 \right) \tag{50}$$

The minimum power consumption, $P_{w,min}$, is derived from the last equation as follows:

$$P_{w,min} = V_{dd} \cdot I_{B,min} = \frac{2 \cdot V_{ov1} \cdot V_{dd} \cdot C_1}{T_{CLK}} \cdot \left(ln\left(\frac{2 \cdot V_{SRo,max}}{\xi}\right) + \frac{V_i}{2 \cdot V_{SRi,max}} - 1 \right) \tag{51}$$

3. Proposed Open-Loop Integrator

As an alternative solution, an open-loop switched-capacitor integrator is presented (Figure 6).

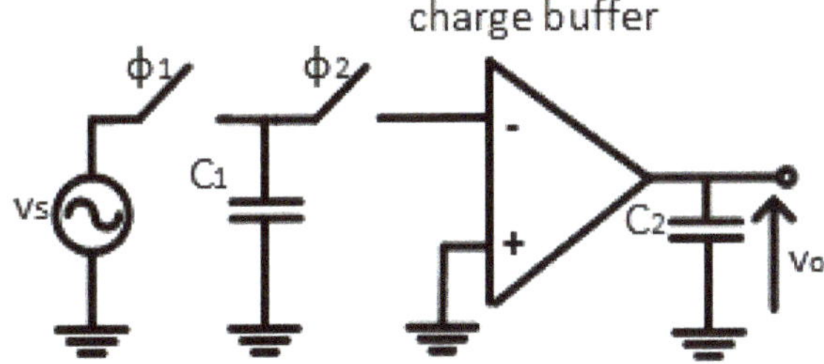

Figure 6. Schematic of the proposed open-loop switched-capacitor integrator.

The active element is an OTA, with a low input impedance, which is called charge buffer. Once the input capacitance, C_1, is connected to the charge buffer input, it is discharged and its charge is transferred to the output capacitance C_2. The proposed open-loop switched-capacitor integrator does not include two input nodes with high and low impedances, unlike the switched-capacitor integrator based on a current conveyor [10,11], but only low impedance input nodes. Therefore, the voltage buffer used at the input in the conveyor integrators is eliminated. These simplifications help to get a more efficient circuit implementation.

According to the operation mode aforementioned, C_1 is connected to the inverting input terminal, it is possible to write:

$$Q_1(n-1) = -Q_2(n) \tag{52}$$

where $Q_1(n-1)$ and $Q_2(n)$ are the charges stored in C_1 and C_2 capacitances, at $n-1$ and n time steps, respectively. From Equation (52), it is obtained:

$$C_1 \cdot v_s(n-1) = -C_2 \cdot v_o(n) \tag{53}$$

where v_o and v_s are the output and the input voltages. Therefore, it is possible to calculate the integrator gain in the Z domain, $\frac{v_o}{v_s}(z)$:

$$\frac{v_o}{v_s}(z) = -\frac{C_1}{C_2} \cdot \frac{z^{-1}}{1 - z^{-1}} \tag{54}$$

By using the proposed approach, we obtain a gain expression, which is identical to the traditional closed-loop integrator reported in Equation (1). In both cases, the desensitization of the gain concerning the OTA parameters is reached as the gain depends only on the C_1 and C_2 capacitor ratio, in the ideal case.

3.1. Small Signal Analysis of the Proposed Charge Buffer

To evaluate the impact of the non-null input resistance and the finite output resistance of the charge buffer, the linear model of the integrator reported in Figure 7 was considered.

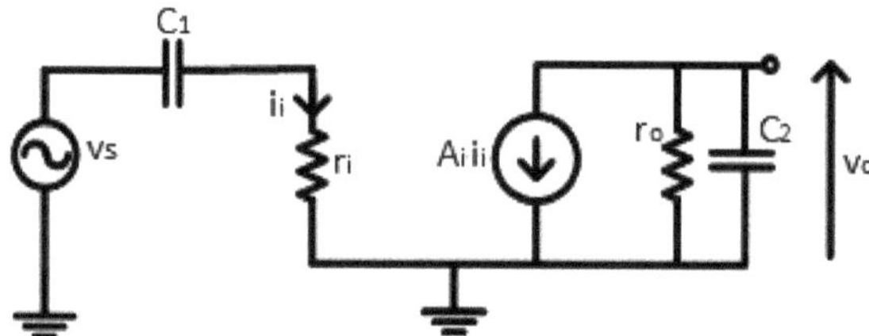

Figure 7. The linear model of the open-loop switched-capacitor integrator.

In practice, the C_1 capacitance is discharged on input resistance r_i, producing the input current i_i. This current is amplified with a current gain A_i by the current amplifier that feds the output load made by the output resistance r_o and the output capacitance C_2.

First of all, the transfer function, $\frac{v_o}{v_s}(s)$, is calculated as previously done for the traditional closed-loop switched-capacitor integrator:

$$\frac{v_o}{v_s}(s) = -A_i \cdot \frac{s \cdot C_1}{1 + s \cdot r_i \cdot C_1} \cdot \frac{r_o}{1 + s \cdot r_o \cdot C_2} \tag{55}$$

Compared to the transfer function of the closed-loop switched-capacitor integrator shown in Equation (3), the transfer function of the open-loop switched-capacitor integrator already is calculated, has an additional pole due to the finite output resistance r_o.

3.2. Transient Analysis of the Proposed Charge Buffer

Assuming a step signal at the input as reported in Equation (2), the output voltage becomes:

$$
\begin{aligned}
v_o(t) &= -\frac{C_1}{C_2} \cdot A_i \cdot V_i \cdot \frac{1}{1 - \frac{r_i}{r_o} \cdot \frac{C_1}{C_2}} \left(e^{-\frac{t}{\tau_{p2}}} - e^{-\frac{t}{\tau_{p1}}} \right) \\
&= -\frac{C_1}{C_2} \cdot A_i \cdot V_i \cdot \frac{1}{1 - \frac{C_1}{A_v \cdot C_2}} \left(e^{-\frac{t}{\tau_{p2}}} - e^{-\frac{t}{\tau_{p1}}} \right)
\end{aligned}
\tag{56}
$$

where:

$$\tau_{p1} = r_i \cdot C_1, \quad \tau_{p2} = r_o \cdot C_2 \qquad A_v = \frac{r_o}{r_i} \tag{57}$$

A_v is the voltage gain. The error on the current gain, A_i, of the current mirror, directly affects the accuracy of the output voltage. This error mainly depends on the transistor mismatch, which can be minimized thanks to the appropriate design of the overdrive of the transistors forming the current mirror [12].

The output resistance r_o, partially drags the charge stored in C_2. Considering a first-order Taylor's expansion for the $e^{-t/\tau_{p2}}$ term, and assuming a unitary current gain, the output voltage, $v_o(t)$, at $t = T_{CLK}/2$, is calculated as follows from Equation (56):

$$v_o\left(\frac{T_{CLK}}{2}\right) \cong -\frac{C_1}{C_2} \cdot \frac{1}{1 - \frac{C_1}{A_v \cdot C_2}} \cdot V_i \cdot \left(1 - \frac{T_{CLK}}{2 \cdot \tau_{p2}} - e^{-\frac{T_{CLK}}{2 \cdot \tau_{p1}}} \right) \tag{58}$$

Three sources of error on the output voltage at $t = \frac{T_{CLK}}{2}$, $v_o\left(\frac{T_{CLK}}{2}\right)$, remain. They are due to:

- finite voltage gain A_v;
- non-null τ_{p1};
- finite τ_{p2}.

The impact of each source of error is evaluated considering the remaining ones disabled.

To evaluate the error, ε_r, due to the finite voltage gain, A_v, it is assumed that τ_{p1} tends to zero and τ_{p2} tends to infinite. In these conditions, the output voltage can be approximated as follows:

$$v_o\left(\frac{T_{CLK}}{2}\right)\Bigg|_{\substack{\tau_{p1} \to 0 \\ \tau_{p2} \to \infty}} \cong -\frac{C_1}{C_2} \cdot \frac{1}{1 - \frac{C_1}{A_v \cdot C_2}} \cdot V_i \tag{59}$$

The corresponding error, ε_r, is calculated as the difference between the ideal voltage obtained using the ideal gain value shown in Equation (54), and the value of the voltage expressed in Equation (59), i.e.,

$$\varepsilon_r = \left| -\frac{C_1}{C_2} \cdot V_i - v_o\left(\frac{T_{CLK}}{2}\right)\Bigg|_{\substack{\tau_{p1} \to 0 \\ \tau_{p2} \to \infty}} \right| = \frac{C_1}{C_2} \cdot \frac{\frac{C_1}{A_v \cdot C_2}}{1 - \frac{C_1}{A_v \cdot C_2}} \cdot V_i \tag{60}$$

To evaluate the error due to τ_{p1}, it is assumed that τ_{p2} tends to infinite. In these conditions, the output voltage can be approximated as follows:

$$v_o\left(\frac{T_{CLK}}{2}\right)\Bigg|_{\tau_{p2} \to \infty} = -\frac{C_1}{C_2} \cdot \frac{1}{1 - \frac{C_1}{A_v \cdot C_2}} \cdot V_i \cdot \left(e^{-\frac{T_{CLK}}{2 \cdot \tau_{p1}}} + 1\right) \tag{61}$$

The error due to τ_{p1}, $\varepsilon_{\tau p1}$, is calculated as the difference between the output voltages expressed in Equations (50) and (52), i.e.:

$$\varepsilon_{\tau p1} = \left| v_o\left(\frac{T_{CLK}}{2}\right)\Bigg|_{\substack{\tau_{p1} \to 0 \\ \tau_{p2} \to \infty}} - v_o\left(\frac{T_{CLK}}{2}\right)\Bigg|_{\tau_{p2} \to \infty} \right| = \frac{C_1}{C_2} \cdot \frac{1}{1 - \frac{C_1}{A_v \cdot C_2}} \cdot V_i e^{\frac{-T_{CLK}}{2 \cdot \tau_{p1}}} \tag{62}$$

The error due to τ_{p2}, $\varepsilon_{\tau p2}$, is calculated as the difference between the output voltages expressed in Equations (58) and (61), i.e.:

$$\varepsilon_{\tau p2} = \left| v_o\left(\frac{T_{CLK}}{2}\right) - v_o\left(\frac{T_{CLK}}{2}\right)\Bigg|_{\tau_{p2} \to \infty} \right| = \frac{C_1}{C_2} \cdot \frac{1}{1 - \frac{C_1}{A_v \cdot C_2}} \cdot V_i \cdot \frac{T_{CLK}}{2 \cdot \tau_{p2}} \tag{63}$$

Figure 8 shows the output voltage behavior.

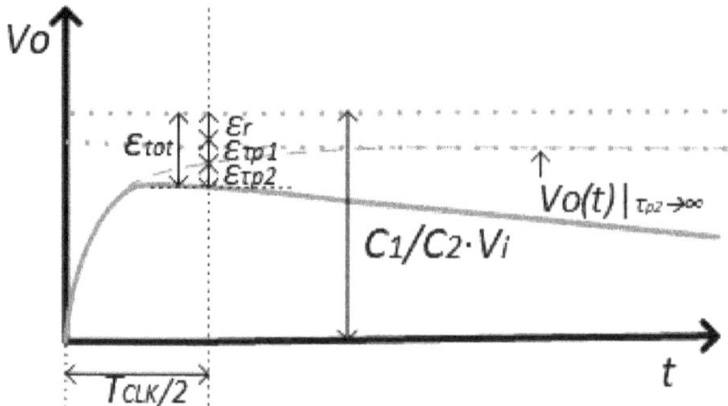

Figure 8. Output voltage behavior of the open-loop switched-capacitor integrator in the linear regime.

The sum of the three error ε_r, $\varepsilon_{\tau p1}$, and $\varepsilon_{\tau p2}$, gives the total error ε_{tot}, which must be less than the required accuracy, ξ:

$$\varepsilon_{tot} = \varepsilon_r + \varepsilon_{\tau p1} + \varepsilon_{\tau p2} < \xi \tag{64}$$

Since the error terms ε_r, $\varepsilon_{\tau p1}$, and $\varepsilon_{\tau p2}$ are positive, each of them must be less than ξ.

3.3. Voltage Gain Requirement of the Proposed Charge Buffer

As calculated in Equation (60), ε_r is less than the ξ, therefore, we obtain

$$A_v > \frac{C_1}{C_2}\cdot\left(1 + \frac{V_i}{\xi}\cdot\frac{C_1}{C_2}\right) \cong \frac{V_i}{\xi}\cdot\frac{C_1^2}{C_2^2} \tag{65}$$

In Equation (65) is very similar to Equation (17), which defines the requirement of the OTA for the closed-loop switched-capacitor integrator. It can be concluded that the charge buffer of the proposed open-loop switched-capacitor integrator requires the same gain of the OTA in the traditional closed-loop solution.

From Equation (58), the sensitivity, $S_{A_v}^{v_0\left(\frac{T_{CLK}}{2}\right)}$, of the output voltage at $\frac{T_{CLK}}{2}$, $v_0\left(\frac{T_{CLK}}{2}\right)$, and A_v is evaluated as follows:

$$S_{A_v}^{v_0\left(\frac{T_{CLK}}{2}\right)} = \frac{A_v}{v_0\left(\frac{T_{CLK}}{2}\right)}\cdot\frac{\partial v_0\left(\frac{T_{CLK}}{2}\right)}{\partial A_v} = \frac{A_v\cdot\frac{C_1}{C_2}}{\left(A_v - \frac{C_1}{C_2}\right)^2} \cong \frac{A_v\cdot\frac{C_1}{C_2}}{\left(A_v - \frac{C_1}{C_2}\right)^2} \tag{66}$$

where the last approximation is valid as $A_v \gg \frac{C_1}{C_2}$. The previous result is very similar to the one obtained for the Cascode OTA in the closed-loop switched-capacitor integrator in Equation (18).

3.4. Input and Output Resistances Requirements of the Proposed Charge Buffer

Assuming $\varepsilon_{\tau p1}$, calculated in Equation (62), less than ξ we obtain

$$\tau_{p1} < \frac{T_{CLK}}{2\cdot ln\left(\dfrac{V_i}{\xi\cdot\left(1-\frac{C_1}{A_v\cdot C_2}\right)\cdot\frac{C_2}{C_1}}\right)} \cong \frac{T_{CLK}}{2\cdot ln\left(\frac{V_i}{\xi}\cdot\frac{C_1}{C_2}\right)} \tag{67}$$

Last approximation in Equation (67) is valid as $A_v \gg \frac{C_1}{C_2}$.

Taking into account the expression of τ_{p1} in Equation (57), the following constraint on the input resistance, r_i, is obtained:

$$r_i < \frac{T_{CLK}}{2 \cdot C_1 \cdot ln\left(\frac{V_i}{\xi} \cdot \frac{C_1}{C_2}\right)} \tag{68}$$

Assuming $\varepsilon_{\tau p2}$, calculated in Equation (63), less than ξ we obtain

$$\tau_{p2} > \frac{T_{CLK}}{2} \cdot \frac{V_i}{\xi \cdot \left(1 - \frac{C_1}{A_v \cdot C_2}\right)} \cdot \frac{C_1}{C_2} \cong \frac{T_{CLK}}{2} \cdot \frac{V_i}{\xi} \cdot \frac{C_1}{C_2} \tag{69}$$

In this case, last approximation is valid as $A_v \gg \frac{C_1}{C_2}$.

Considering τ_{p2} in Equation (57), from Equation (69) it is derived the following constraint on the output resistance, r_0:

$$r_0 > \frac{T_{CLK}}{2} \cdot \frac{V_i}{\xi} \cdot \frac{C_1}{C_2^2} \tag{70}$$

As already done for the voltage gain, A_v, from Equation (58) the sensitivity, $S_{r_i}^{v_0\left(\frac{T_{CLK}}{2}\right)}$, of the output voltage at $\frac{T_{CLK}}{2}$, $v_0\left(\frac{T_{CLK}}{2}\right)$, and r_i is evaluated as follows:

$$S_{r_i}^{v_0\left(\frac{T_{CLK}}{2}\right)} = \frac{r_i}{v_0\left(\frac{T_{CLK}}{2}\right)} \cdot \frac{\partial v_0\left(\frac{T_{CLK}}{2}\right)}{\partial r_i} = \frac{T_{CLK}}{2 \cdot \tau_{p1}} \cdot \frac{e^{\frac{-T_{CLK}}{2\tau_{p1}}}}{1 - \frac{T_{CLK}}{2 \cdot \tau_{p2}} - e^{\frac{-T_{CLK}}{2 \cdot \tau_{p1}}}} \cong \frac{T_{CLK}}{2 \cdot \tau_{p1}} e^{\frac{-T_{CLK}}{2 \cdot \tau_{p1}}} \tag{71}$$

where last approximation is valid as $\frac{T_{CLK}}{2 \cdot \tau_{p2}} - e^{\frac{-T_{CLK}}{2 \cdot \tau_{p1}}} \ll 1$. This inequation is verified as the condition imposed by Equations (67) and (69) are satisfied, since, generally, $V_i \gg \xi$ and $\frac{C_1}{C_2} \geq 1$. It can be seen that the result of the calculation of the sensitivity of the output voltage compared to r_i is very similar to the one obtained for the calculation of the output voltage sensitivity for g_m of the Cascode OTA in the closed-loop switched-capacitor integrator in Equation (22).

Regarding the sensitivity, it can be concluded that the proposed switched-capacitor integrator, despite working in an open-loop configuration, has a robustness to PVT variations similar to the closed-loop switched-capacitor integrator.

However, since the performance of the proposed open-loop switched-capacitor integrator depends on the output resistance of the charge buffer, r_0, the sensitivity, $S_{r_0}^{v_0\left(\frac{T_{CLK}}{2}\right)}$, of the output voltage at $\frac{T_{CLK}}{2}$, $v_0\left(\frac{T_{CLK}}{2}\right)$, and r_0 is calculated as:

$$S_{r_0}^{v_0\left(\frac{T_{CLK}}{2}\right)} = \frac{r_i}{v_0\left(\frac{T_{CLK}}{2}\right)} \cdot \frac{\partial v_0\left(\frac{T_{CLK}}{2}\right)}{\partial r_i} = \frac{T_{CLK}}{2 \cdot \tau_{p2}} \cdot \frac{1}{1 - \frac{T_{CLK}}{2 \cdot \tau_{p2}} - e^{\frac{-T_{CLK}}{2 \cdot \tau_{p2}}}} \cong \frac{T_{CLK}}{2 \cdot \tau_{p2}} \tag{72}$$

According to Equation (69) and considering that $\frac{V_i}{\xi} \gg 1$ and $\frac{C_1}{C_2} \geq 1$, it can be assumed that $S_{r_0}^{v_0\left(\frac{T_{CLK}}{2}\right)}$ is quite less than 1. Therefore, the impact of the r_0 variation on the integrator performance is limited.

3.5. Circuit Implementation of the Proposed Charge Buffer

Figure 9 shows a possible circuit implementation of the charge buffer. The switched capacitors network at the output nodes is used to set the output common-mode voltage at V_{cm}. Reference V_{b1} is designed to set the input common-mode voltage at V_{cm}. To keep M_2 and M_8 transistors in the saturation region, their source-drain voltage must be more than their saturation voltage, i.e.:

$$V_{SD2} = V_{dd} - V_{cm} > V_{SD2,sat} \tag{73}$$

It is supposed that V_{cm} is equal to half supply voltage and the saturation voltages correspond to the transistor overdrive, V_{ov2}, from Equation (73) we derive

$$V_{ov2} < \frac{V_{dd}}{2} \tag{74}$$

To bias the M_1 transistor in the saturation region, it must be guaranteed that its source-drain voltage, V_{SD1}, overcomes its saturation voltage, corresponding to the transistor overdrive, V_{ov1}, i.e.:

$$V_{SD1} > V_{ov1} \tag{75}$$

From the last equation, we obtain

$$V_{SD1} = V_{cm} - V_{dd} + |V_{THP}| + V_{ov2} > V_{ov1} \tag{76}$$

Assuming $V_{cm} = V_{dd}/2$, the last equation can be rearranged to obtain a constraint on the difference between the M_1 and M_2 transistors overdrives, ΔV_{ov2-1}, i.e.:

$$\Delta V_{ov2-1} = V_{ov2} - V_{ov1} > \frac{V_{dd}}{2} - |V_{THP}| \tag{77}$$

Using the finFET 16 nm we obtain the result $V_{dd} = 0.95$ V, $V_{THP} = 0.4$ V. Consequently, ΔV_{ov2-1} must be higher than 75 mV.

According to Equations (74) and (77), using a charge buffer in open-loop configuration gives more flexibility to the design since larger overdrives can be defined for the transistors, to employing an OTA in a closed-loop fashion. This is extremely important at low voltage supply.

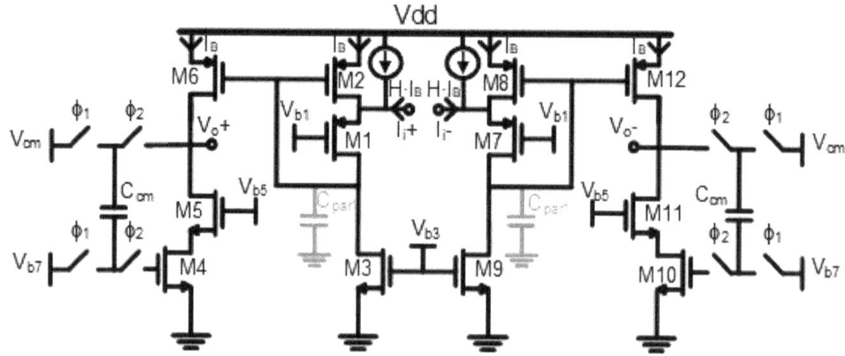

Figure 9. Circuit implementation of the charge buffer.

The input and the output resistances, r_i and r_o, are calculated as follows:

$$r_i \cong \frac{1}{g_{m1} \cdot g_{m2} \cdot r_{ds3}} = \frac{V_{ov1} \cdot V_{ov2}}{4 \cdot V_{A3} \cdot I_B}; \quad r_o \cong r_{ds6} = \frac{V_{A6}}{I_B} \tag{78}$$

where g_{m1} and g_{m2} are the transconductance of M_1 and M_2 transistors, respectively, while r_{ds3} and r_{ds6} are the output resistance of M_3 and M_6 transistors, respectively.

The voltage gain, A_v, is calculated as follows:

$$A_v = \frac{r_o}{r_i} = \frac{4 \cdot V_{A3} \cdot V_{A6}}{V_{ov1} \cdot V_{ov2}} \tag{79}$$

The telescopic Cascode OTA reported in Figure 4 implements the boost of the output resistance, r_o. On the other end, the proposed charge buffer circuit enables the boosting of the transconductance of the input transistors, g_{m1}, by a $g_{m2} \cdot r_{ds3}$ factor, lowering the input resistance, r_i. The impact on the final voltage gain is similar as demonstrated by the similitude of the voltage gain expressions reported in Equations (79) and (27), even if the voltage gain, A_v, of the proposed charge buffer results the double with respect to the telescopic Cascode OTA.

In both cases, the power consumption is determined by the settling requirements, i.e., both the time constants τ_p and τ_{p1} for the closed-loop and the proposed open-loop switched-capacitor integrator, respectively. The time constant, τ_p, depends on $1/g_{m1}$. In this case, the only possibility to increase g_{m1} is to increase the bias current I_B of the telescopic Cascode OTA, since the input transistor overdrive is bound to bias constraints. For the proposed integrator, the time constant τ_{p1} is proportional to r_i. However, in the last case, as the boost on the input transistor transconductance lowers the input resistance, r_i, a significant power saving is obtained.

3.6. Slew-Rate Analysis of the Proposed Charge Buffer

According to Equation (56), the maximum rate of variation of the output voltage, i.e., the slew-rate, is obtained at the initial instant, $t = 0$ s:

$$SR = \left. \frac{dv_o(t)}{dt} \right|_{max} = \frac{C_1}{C_2} \cdot \frac{1}{1 - \frac{C_1}{A_v \cdot C_2}} \cdot V_{SRi,max} \cdot \left(\frac{1}{\tau_{p1}} - \frac{1}{\tau_{p2}} \right) \cong \frac{V_{SRo,max}}{\tau_{p1}} \tag{80}$$

where:

$$V_{SRo,max} = \frac{C_1}{C_2} \cdot \frac{1}{1 - \frac{C_1}{A_v C_2}} \cdot V_{SRi,max} \cong \frac{C_1}{C_2} \cdot V_{SRi,max} \tag{81}$$

The last approximation in Equation (81) is valid assuming $\tau_{p1} \ll \tau_{p2}$.

The slew-rate depends on the bias current I_B and the output capacitance C_2, i.e.:

$$SR = \frac{I_B}{C_2} \tag{82}$$

where I_B is the bias current of each branches composing the charge buffer drawn in Figure 9.

Combining Equations (80) and (82) and considering the expression of τ_{p1} and r_i reported in Equations (57) and (78), respectively, we derive:

$$V_{SRo,max} = \frac{I_B}{C_2} \cdot \tau_{p1} = \frac{C_1}{C_2} \cdot \frac{V_{ov1} \cdot V_{ov2}}{4 \cdot V_{A3}} \tag{83}$$

where V_{ov1} and V_{ov2} are the overdrive voltage of M_1 and M_2 transistors, respectively, and V_{A3} is the early voltage of M_3 transistor.

Combining Equations (81) and (84), the value of $V_{SRi,max}$ is obtained:

$$V_{SRi,max} = \frac{V_{ov1} \cdot V_{ov2}}{4 \cdot V_{A3}} \cdot \frac{1}{1 - \frac{C_1}{A_v \cdot C_2}} \cong \frac{V_{ov1} \cdot V_{ov2}}{4 \cdot V_{A3}} \tag{84}$$

The output voltage range where the charge buffer operates in the linear regime, $V_{SRo,max}$, has been reduced by a factor equal to $\frac{4 \cdot V_{A3}}{V_{ov2}}$ concerning the traditional closed-loop switched-capacitor integrator with the OTA.

Due to its differential structure, the charge buffer starts slewing as $V_i > 2 \cdot V_{SRi,max}$. If a slewing period is considered, the differential output voltage, $v_{od}(t)$, can be calculated as follows:

$$\begin{cases} v_{od}(t) = -2 \cdot \dfrac{V_{SRo,max}}{\tau_{p1}} \cdot t \quad for \ 0 < t < \tau_s \\ v_{od}(t) = \left(-\dfrac{C_1}{C_2} \cdot \dfrac{1}{1 - \frac{C_1}{A_v \cdot C_2}} \cdot V_i + 2 \cdot V_{SRo,max}\right)\left(1 - \dfrac{t - \tau_s}{\tau_{p2}}\right) - 2 \cdot V_{SRo,max} \cdot \left(1 - \dfrac{t - \tau_s}{\tau_{p2}} - e^{-\frac{t - \tau_s}{\tau_{p1}}}\right) for \ t > \tau_s \end{cases}$$
(85)

where τ_s is the duration of the slewing period. The charge buffer slews until the output differential voltage, $v_{od}(t)$, is less than $2 \cdot V_{SRo,max}$ concerning the final value in steady-state, neglecting the losses due to the output resistance (i.e., $\tau_{p2} \to \infty$):

$$-2 \cdot \frac{V_{SRo,max}}{\tau_{p1}} \cdot \tau_s = -\frac{C_1}{C_2} \cdot \frac{1}{1 - \frac{C_1}{A_v \cdot C_2}} \cdot V_i + 2 \cdot V_{SRo,max} \cong -\frac{C_1}{C_2} \cdot V_i + 2 \cdot V_{SRo,max}$$
(86)

From the combination of the last equation and Equation (83), the expression of τ_s is derived:

$$\tau_s = \tau_{p1} \cdot \left(\frac{C_1}{C_2} \cdot \frac{V_i}{2 \cdot V_{SRo,max}} - 1\right) = \tau_{p1} \cdot \left(\frac{V_i}{2 \cdot V_{SRi,max}} - 1\right)$$
(87)

3.7. SNR Analysis of the Proposed Open-Loop Switched-Capacitor Integrator

By focusing the analysis on a single branch, V_{o+} must satisfy the following inequation to guarantee that M_4 transistors operate in saturation region:

$$V_{o+} > V_{DS,sat4} + V_{S4} = V_{ov} + V_{S4}$$
(88)

where V_{S4} and $V_{DS,sat4}$ are the source and the saturation voltages of M_4 transistor, respectively. It is assumed that $V_{ds,sat4}$ is equal to V_{ov}. The bias voltage V_{b5} was chosen to make M_5 transistor operating in saturation, i.e.,

$$V_{ds5} = V_{S4} > V_{DS,sat5} = V_{ov}$$
(89)

where V_{DS5}, and $V_{DS,sat5}$ are the drain-source, and the saturation voltages of M_4 transistor, respectively. In this case, it is assumed that $V_{ds,sat1}$ is equal to V_{ov}. V_{S1} is derived from the V_{b5} bias voltage, by dropping the gate-drain voltage of the M_4 transistors, V_{GS4}, i.e.:

$$V_{S4} = V_{b1} - V_{GS4} = V_{b1} - V_{TH} - V_{ov}$$
(90)

V_{b1} can be designed to make V_{S4}, and, hence, V_{DS5} equal to $V_{DS5,sat}$, i.e., V_{ov}. If so, from Equation (88) we derive the minimum output voltage, $V_{o+,min}$:

$$V_{o+} > V_{ov} + V_{DS5,sat} = 2 \cdot V_{ov} = V_{o+,min}$$
(91)

As the output voltage starts swinging from the common-mode voltage, V_{cm}, down to $V_{o,min}$, it is possible to calculate the maximum output voltage swing, V_{swing}:

$$V_{swing} = 2 \cdot (V_{cm} - V_{o+,min}) = 2 \cdot \left(\frac{V_{dd}}{2} - 2 \cdot V_{ov}\right) = V_{dd} - 4 \cdot V_{ov}$$
(92)

where the 2 factor is due to the differential architecture. It is assumed that V_{cm} is equal to half V_{dd}.

Assuming that the thermal noise due to the switches around C_1, $2 \cdot \frac{KT}{C_1}$ is dominant, from Equation (92), the signal to noise ratio of the overall open loop switched capacitor integrator, SNR_{OL}, is calculated as follows:

$$SNR_{OL} = \frac{1}{2} \cdot \frac{V_{swing}^2}{v_{n,o}^2} = \frac{1}{2} \cdot \frac{(V_{dd} - 4 \cdot V_{ov})^2}{2 \cdot \frac{K \cdot T}{C_1} \cdot \left(\frac{C_1}{C_2}\right)^2} = \frac{(V_{dd} - 4 \cdot V_{ov})^2 \cdot C_2^2}{4 \cdot K \cdot T \cdot C_1} \tag{93}$$

where $\overline{v_{n,o}^2}$ is the total output noise, which is given by the thermal noise contribution due to the C_1 switched-capacitor multiplied by the square of the integrator gain $\left(\frac{C_1}{C_2}\right)^2$, and the $\frac{1}{2}$ factor takes into account that the input signal is a sinusoid.

The resulting SNR_{OL} is slightly higher than the SNR_{CL} calculated by Equation (43). The output noise is about the same since the noise contribution of C_1 is assumed dominant. However, assuming $V_{TH} = 0.275$ V and $V_{dd} = 0.95$ V as for the finFET technology and $V_{ov} = 0.1$ V, due to bias constraint as defined by Equation (43), the output voltage swing for the proposed switched-capacitor integrator is higher.

3.8. Power Consumption Requirement of the Proposed Charge Buffer

The minimum power consumption, $P_{w,min}$, is given by the product of the supply voltage V_{dd}, by the minimum total bias current $I_{BTOT.min}$:

$$P_{w,min} = V_{dd} \cdot I_{BTOT,min} \cdot 4 \cdot \left(1 + \frac{H}{2}\right) \cdot V_{dd} \cdot I_{B,min} \tag{94}$$

where $I_{B,min}$ is the minimum I_B bias current. The power requirement is calculated according to the settling time requirement. In practice, the minimum bias current $I_{B,min}$ is derived assuming that the $\varepsilon_{\tau p1}$ error must be less than the required accuracy ξ, i.e.:

$$\varepsilon_{\tau p1} < \xi \tag{95}$$

As $V_i < 2 \cdot V_{SRi,max}$, the charge buffer is in the linear regime, where the expression of $I_{B,min}$ is derived from Equation (62):

$$I_{B,min} = \frac{2 \cdot C_1 \cdot V_{SRi,max}}{T_{CLK}} \cdot ln\left(\frac{C_1}{C_2} \cdot \frac{V_i}{\xi}\right) \tag{96}$$

Combining Equations (94) and (96), the minimum required power consumption, $P_{w,min}$, is calculated as follows:

$$P_{w,min} = \frac{8 \cdot V_{dd} \cdot \left(1 + \frac{H}{2}\right) \cdot C_1 \cdot V_{SRi,max}}{T_{CLK}} \cdot ln\left(\frac{C_1}{C_2} \cdot \frac{V_i}{\xi}\right) \tag{97}$$

As $V_i > 2 \cdot V_{SRi,max}$, the charge buffer starts slewing. Considering the expression of the differential output voltage $v_{od}(t)$, including the slewing period reported in Equation (85), the calculation of the error due to τ_{p1}, $\varepsilon_{\tau p1}$, is updated as follows:

$$\varepsilon_{\tau p1} = \left| v_0\left(\frac{T_{CLK}}{2}\right)\Big|_{\substack{\tau_{p1} \to 0 \\ \tau_{p2} \to \infty}} - v_0\left(\frac{T_{CLK}}{2}\right)\Big|_{\tau_{p2} \to \infty} \right| = 2 \cdot V_{SRo,max} \cdot e^{-\frac{\frac{T_{CLK}}{2} - \tau_s}{\tau_{p1}}} \tag{98}$$

Considering $\varepsilon_{\tau p1}$ as reported in the previous equation, the constraint on τ_{p1} is derived from Equation (95):

$$\tau_{p1} < \frac{T_{CLK}}{2\left(ln\left(\frac{2 \cdot V_{SRo,max}}{\xi}\right) + \frac{V_i}{2 \cdot V_{SRi,max}} - 1\right)} \tag{99}$$

Looking at τ_{p1} and r_i in Equations (57) and (78), respectively, the minimum bias current that satisfies Equation (99), $I_{B,min}$ is calculated as follows:

$$I_{B,min} = \frac{2{\cdot}C_1{\cdot}V_{SRi,max}}{T_{CLK}}{\cdot}\left(ln\left(\frac{2{\cdot}V_{SRo,max}}{\xi}\right) + \frac{V_i}{2{\cdot}V_{SRi,max}} - 1\right) \tag{100}$$

Combining Equations (97) and (100), the minimum required power consumption is calculated as follows:

$$P_{w,min} = \frac{8{\cdot}V_{dd}{\cdot}C_1{\cdot}\left(1 + \frac{H}{2}\right){\cdot}V_{SRi,max}}{T_{CLK}}{\cdot}\left(ln\left(\frac{2{\cdot}V_{SRo,max}}{\xi}\right) + \frac{V_i}{2{\cdot}V_{SRi,max}} - 1\right) \tag{101}$$

Figure 10 shows the power consumption of the proposed switched-capacitor integrator and the closed-loop switched-capacitor integrator plotted as a function of V_i.

The two curves in Figure 10 are obtained plotting the Equations (97) and (101) for the proposed design, and Equations (46) and (51) for the closed-loop switched-capacitor integrator. The common design parameters are reported in Table 1.

Table 1. Design parameters of the switched capacitor integrator.

Design Parameter	Value
V_{ov}	0.1 V
V_{ov2}	0.2 V
V_{dd}	0.95 V
V_i	31.25 mV
ξ (accuracy)	1 mV
T_{CLK}	7 ns
C_1	1.25 pF
C_2	312.5 fF
C_1/C_2	4
C_{par1}	27 fF
V_{A3}	1 V
V_{A6}	18 V

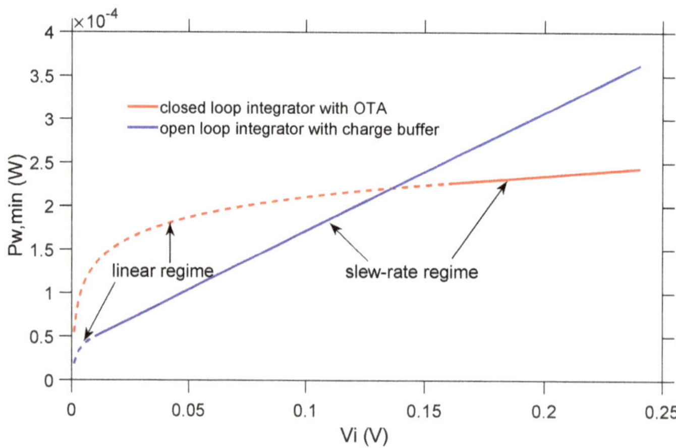

Figure 10. Minimum power consumption for the closed and the open-loop switched-capacitor integrators.

The transistors overdrives have been defined according to the constraints derived from Equations (25), (74), and (77). The values of C_{par1}, $V_{A3,}$ and V_{A6} are estimated from the simulation results. The H factor was set to 2 for the proposed design.

The minimum power required by the closed-loop switched-capacitor integrator is higher than the proposed open loop integrator for an input signal up to 140 mV large. For $V_i = 31.25$ mV, the proposed circuit requires a minimum power of 76 μW, while the closed-loop switched-capacitor integrator requires about 173 μW, i.e., more than the double.

3.9. Small Signal Analysis of the Charge Buffer Considering the Parasitic Capacitance C_{par1}

The small-signal equivalent circuit shown in Figure 7 is a first-order approximation of the small-signal behavior of the proposed transistor-level open-loop switched-capacitor integrator. Considering also the C_{par1} as a parasitic capacitance shown in Figure 9, a more accurate transfer function is obtained:

$$\frac{v_o}{v_s} = -\frac{s \cdot C_1 \cdot \left(1 + \frac{s \cdot C_{par1}}{g_{m2}}\right)}{1 + s \cdot \left(\frac{C_{par1}}{g_{m2}} + \frac{C_1}{g_{m1} \cdot g_{m2} \cdot r_{ds3}}\right) + s^2 \cdot \frac{C_1 \cdot C_{par1}}{g_{m1} \cdot g_{m2}}} \cdot \frac{r_o}{1 + s \cdot r_o \cdot C_2} \tag{102}$$

where hr_{ds3} is the output resistances of M_3 transistor. The parasitic capacitance, C_{par1}, mainly depends on the gate capacitances of M_2 and M_6 transistors. Therefore, it can be approximated as follows:

$$C_{par1} \cong \frac{2}{3} \cdot W_2 \cdot L_2 \cdot C_{ox} + \frac{2}{3} \cdot W_6 \cdot L_6 \cdot C_{ox} = \frac{4}{3} \cdot W_2 \cdot L_2 \cdot C_{ox} \tag{103}$$

where W_2 and L_2, and W_6 and L_6, are the width and the length of M_2 and M_6 transistors.

Concerning the transfer function reported in Equation (55), the transfer function in Equation (84) includes a further zero, z_1:

$$z_1 = -\frac{g_{m2}}{C_{par1}} \tag{104}$$

This zero is considered to be at a very high frequency and it does not produce significant effects on the step response of the proposed circuit.

Moreover, two complex poles appear in the transfer function. Their frequency, ω_o, and quality factor, Q, are calculated as follows:

$$\omega_0 = \sqrt{\frac{g_{m1} \cdot g_{m2}}{C_1 \cdot C_{par1}}} = 2 \cdot I_B \cdot \sqrt{\frac{H+1}{V_{ov1} V_{ov2} C_1 C_{par1}}} \quad \text{and}$$

$$Q = \frac{1}{\sqrt{\frac{g_{m1} \cdot C_{par1}}{g_{m2} \cdot C_1}} + \sqrt{\frac{C_1}{r_{ds3}^2 \cdot g_{m1} \cdot g_{m2} \cdot C_{par1}}}} = \frac{1}{\left(\sqrt{H+1}\right) \cdot \left(\sqrt{\frac{V_{ov2} \cdot C_{par1}}{V_{ov1} \cdot C_1}} + \sqrt{\frac{V_{ov1} \cdot V_{ov2} \cdot C_1}{4 \cdot V_{A3}^2 \cdot C_{par1}}}\right)} \tag{105}$$

A high Q factor determines a large overshoot, OS, on the step response of the proposed circuit, and wide oscillations, which can have a severe impact on the accuracy of the output voltage. Otherwise, as the circuit is excessively dumped, the step response slows significantly. The criterion here adopted is to limit the overshoot to the required accuracy, ξ, i.e.:

$$OS = \frac{C_1}{C_2} \cdot V_i \cdot e^{\frac{-\frac{\pi}{2 \cdot Q}}{\sqrt{1 - \frac{1}{4 \cdot Q^2}}}} = \xi \tag{106}$$

The previous equation is valid in linear regime; otherwise, in case of slewing of the charge buffer, the overshot is calculated as follows:

$$OS = V_{SRo,max} \cdot e^{\frac{-\frac{\pi}{2 \cdot Q}}{\sqrt{1 - \frac{1}{4 \cdot Q^2}}}} = \xi \tag{107}$$

For the proposed design, the last equation is satisfied for a Q value of about to 0.75. The Q factor can be reduced by operating on V_{ov2} and V_{ov1}, or, by acting on the H factor, which gives a further degree of freedom to the design.

The desired value of Q is reached by designing an H factor of 2.

4. Simulation Results

A transistor-level design of the proposed switched-capacitor circuit was performed in finFET 16 nm CMOS technology. The design parameter reported in Table 1 were considered. The input signal, V_i, was assumed equal to 31.25 mV. The bias current, I_B, set to 10 µA, corresponds to the minimum value, $I_{B,min}$, as predicted by Equation (100). The H factor was set to 2 as derived from Equation (107). The minimum power consumption of the core circuit was 76 µA, as predicted by Equation (83).

According to Equation (65), the required voltage gain is 56 dB, while a voltage gain of 71 dB results from simulations. Similarly, the required output resistance obtained from Equation (70) was 1.4 MΩ, while the value obtained through simulations was 1.85 MΩ. Therefore, we can conclude that the voltage gain and the output resistance requirements were largely satisfied.

Figure 11 shows the response of the circuit to an input step of 31.25 mV for the theoretical model and simulations. The two curves are very close, proving the validity of the proposed circuit model. Based on the design parameters, the expected error on the output voltage at $T_{CLK}/2$ was 1 mV. This results from both the model prediction and the simulations.

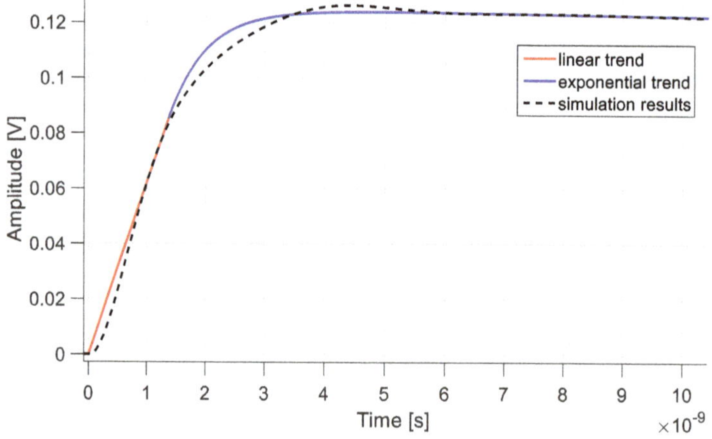

Figure 11. Simulated and predicted step response of the proposed open-loop switched-capacitor integrator.

The simulation results show a slightly marked overshot due to the complex poles generated by the internal loop including M_1 and M_2 transistors, as predicted in paragraph 3.9. However, the first-order model gives a valid approximation of the circuit behavior especially in the steady-state regime.

Table 2 summarizes the required values of the design parameters and their values obtained through simulations.

Table 2. Required and simulated design parameters for the open-loop switched-capacitor integrator.

Design Parameter	Required Value	Simulated Value
A_v	>56 dB	71 dB
r_o	>1.4 MΩ	1.85 MΩ
$I_{B,min}$	10 µA	10 µA
ξ (*accuracy*)	1 mV	1 mV

Table 3 reports the performance summary of the proposed switched-capacitor integrator and compares it to the state-of-the-art approach. The following figure of merit (*FoM*) is introduced for a fast comparison

$$FoM \;=\; \frac{SNR}{\frac{P_w}{N} \cdot \frac{f_{CLK}}{2 \cdot OSR}} \tag{108}$$

where N is the number of poles of the switched-capacitor filter under consideration, P_w is its power, f_{CLK} is the clock frequency and OSR is the oversampling ratio, i.e., the ratio between half clock frequency and the maximum signal bandwidth.

Table 3. Performance summary and state-of-the-art comparison.

Parameter	This work	[13]	[14]
Supply	0.95 V	1.8 V	1.2 V
Power	76 μW	4.3 mW	8.4 mW
Technology	16 nm FinFET	180 nm	65 nm
Number of poles	1	4	12
f_{CLK}	143 MHz	300 MHz	430 MHz
OSR	1.57	11.3	9.23
SNR	61.5 dB	74.9 dB	45 dB
FoM	179.3 dB	175.8 dB	150.6 dB

As can be seen in Table 3, the proposed work is well compared to the state of the art in terms of *FoM*.

5. Conclusions

An architecture of a switched-capacitor integrator including a charge buffer operating in open-loop have been proposed and designed in finFET 16 nm technology. As for the switched capacitor filters, the gain of the proposed integrator is given by the capacitor ratio, guarantying desensitization concerning the PVT variations. Furthermore, the proposed circuit is more suitable for low voltage supplies. Moreover, the analytical study demonstrated that the proposed integrator is more efficient than the traditional closed-loop switched-capacitor integrator for input signal amplitude less than 140 mV. The proposed switched-capacitor integrator results were more than twice the efficiency when compared to the traditional closed-loop switched-capacitor filter, as it consumes 76 μW from the 0.95 V supply, assuming an input voltage of 31.25 mV and a clock period of 7 ns. The proposed work results were satisfactory when compared to the state-of-the-art in terms of the figure of merit.

Author Contributions: Conceptualization, S.D. and S.M.; Methodology, S.D., S.M., P.P. and M.B.; Software, S.D. and S.M.; Validation, S.D., S.M., P.P. and M.B.; Formal Analysis, S.D.; Investigation, S.D., S.M., P.P. and M.B.; Resources, S.D., S.M., P.P. and M.B.; Data Curation, S.D. and S.M.; Writing—Original Draft Preparation, S.D.; Writing—Review & Editing, S.D., S.M., P.P. and M.B.; Supervision, S.D. All authors have read and agreed to the published version of the manuscript.

Funding: This research received no external funding.

Conflicts of Interest: The authors declare no conflict of interest.

References

1. Zhang, X.; Chan, P.K. A Low-Power Switched-Capacitor Capacitive Transducer With High Resolution. *IEEE Trans. Instrum. Meas.* **2008**, *57*, 1492–1499. [CrossRef]
2. Woo, S.; Cho, J.-K. A Switched-Capacitor Filter With Reduced Sensitivity to Reference Noise for Audio-Band Sigma–Delta D/A Converters. *IEEE Trans. Circuits Syst. II Express Brief* **2016**, *63*, 361–365. [CrossRef]
3. Xu, Y.; Kinget, P.R. A Switched-Capacitor RF Front End With Embedded Programmable High-Order Filtering. *IEEE J. Solid State Circuits* **2016**, *51*, 1154–1167. [CrossRef]
4. Murmann, B.; Boser, B.E. A 12-bit 75-MS/s Pipelined ADC Using Open-Loop Residue Amplification. *IEEE J. Solid State Circuits* **2003**, *38*, 2040–2050. [CrossRef]

5.   Quinn, P.J.; van Roermund, A.H.M. *Switched-Capacitor Techniques for High-Accuracy Filter and ADC Design*; Springer: New York, NY, USA, 2007; p. 37.

6.   Maloberti, F. *Analog Design for CMOS VLSI Systems*; Kluwer Academic: Alphen aan den Rijn, The Netherlands, 2001; p. 115.

7.   Chauhan, D.Y.S.; Lu, D.D.; Sriramkumar, V.; Khandelwal, S.; Duarte, J.P.; Payvadosi, N.; Niknejad, A.; Hu, C. *FinFET Modeling for IC Simulation and Design: Using the BSIM-CMG Standard*; Elsevier: Amsterdam, The Netherlands, 2015; p. 26.

8.   Flandre, D.; Viviani, A.; Eggermont, J.-P.; Gentinne, B.; Jespers, P.G.A. Improved Synthesis of Gain-Boosted Regulated-Cascode CMOS Stages Using Symbolic Analysis and gm/ID Methodology. *IEEE J. Solid State Circuits* **1997**, *32*, 1006–1012. [CrossRef]

9.   De Langen, K.-J.; Huijsing, J.H. Compact Low-Voltage Power-Efficient Operational Amplifier Cells for VLSI. *IEEE J. Solid State Circuits* **1998**, *33*, 1492–1496. [CrossRef]

10.   Kaabi, H.; Motlagh, M.R.J.; Ayatollahi, A. A novel current-conveyor-based switched-capacitor integrator. In Proceedings of the 2005 IEEE International Symposium on Circuits and Systems, (ISCAS 2005), Kobe, Japan, 23–26 May 2005.

11.   Balasubramaniam, H.; Hofmann, K. Robust design methodology for switched capacitor delta sigma modulators based on current conveyors. In Proceedings of the 2014 IEEE 12th International New Circuits and Systems Conference (NEWCAS 2014), Trois-Rivieres, QC, Canada, 22–25 June 2014.

12.   Albiol, M.; González, J.L.; Alarcón, E. Mismatch and Dynamic Modeling of Current Sources in Current-Steering CMOS D/A Converters: An Extended Design Procedure. *IEEE Trans. Circuits Syst. I Regul. Pap.* **2004**, *51*, 159–169. [CrossRef]

13.   Payandehnia, P.; Maghami, H.; Mirzaie, H.; Kareppagoudr, M.; Dey, S.; Tohidian, M.; Temes, G.C. A 0.49–13.3 MHz Tunable Fourth-Order LPF with Complex Poles Achieving 28.7 dBm OIP3. *IEEE Trans. Circuits Syst. I Regul. Pap.* **2018**, *65*, 2353–2364. [CrossRef]

14.   Huang, M.F.; Kuo, M.C.; Yang, T.Y.; Huang, X.L. A 58.9-dB ACR, 85.5-dB SBA, 5–26-MHz Configurable-Bandwidth, Charge-Domain Filter in 65-nm CMOS. *IEEE J. Solid State Circuits* **2013**, *48*, 2827–2838. [CrossRef]

 *electronics*

Article

10-GHz Fully Differential Sallen–Key Lowpass Biquad Filters in 55nm SiGe BiCMOS Technology

Francesco Centurelli, Pietro Monsurrò, Giuseppe Scotti *, Pasquale Tommasino and Alessandro Trifiletti

Dipartimento di Ingegneria dell'Informazione, Elettronica e Telecomunicazioni (DIET), University of Rome "La Sapienza", 00185 Roma, Italy; francesco.centurelli@uniroma1.it (F.C.); pietro.monsurro@uniroma1.it (P.M.); pasquale.tommasino@uniroma1.it (P.T.); alessandro.trifiletti@uniroma1.it (A.T.)
* Correspondence: giuseppe.scotti@uniroma1.it; Tel.: +39-0644585690

Received: 15 February 2020; Accepted: 26 March 2020; Published: 28 March 2020

Abstract: Multi-GHz lowpass filters are key components for many RF applications and are required for the implementation of integrated high-speed analog-to-digital and digital-to-analog converters and optical communication systems. In the last two decades, integrated filters in the Multi-GHz range have been implemented using III-V or SiGe technologies. In all cases in which the size of passive components is a concern, inductorless designs are preferred. Furthermore, due to the recent development of high-speed and high-resolution data converters, highly linear multi-GHz filters are required more and more. Classical open loop topologies are not able to achieve high linearity, and closed loop filters are preferred in all applications where linearity is a key requirement. In this work, we present a fully differential BiCMOS implementation of the classical Sallen Key filter, which is able to operate up to about 10 GHz by exploiting both the bipolar and MOS transistors of a commercial 55-nm BiCMOS technology. The layout of the biquad filter has been implemented, and the results of post-layout simulations are reported. The biquad stage exhibits excellent SFDR (64 dB) and dynamic range (about 50 dB) due to the closed loop operation, and good power efficiency (0.94 pW/Hz/pole) with respect to comparable active inductorless lowpass filters reported in the literature. Moreover, unlike other filters, it exploits the different active devices offered by commercial SiGe BiCMOS technologies. Parametric and Monte Carlo simulations are also included to assess the robustness of the proposed biquad filter against PVT and mismatch variations.

Keywords: active filters; anti-aliasing filters; HBT; inductorless; low-pass filters; SiGe

1. Introduction

Integrated multi-GHz-band lowpass filters are required as antialiasing filters for very high-speed analog-to-digital (ADC) and digital-to-analog (DAC) converters [1] in applications such as wideband spectrum monitoring, high bit-rate optical communications [2,3] and wideband measurement systems [4,5]. They have to be designed in silicon technology to be integrated on the same chip with the converter blocks, thus minimizing off-chip interfaces, and should possibly avoid the use of spiral inductors, to minimize chip area. The main performance requirements are related to off-band suppression, that forces the use of high order filters, and linearity, that should be better than that of the ADC/DAC, not to limit the overall system performance. A fully differential approach is typically required, to desensitize from common-mode disturbances, reduce even-order harmonics and improve the signal-to-noise ratio (SNR).

Inductorless GHz-band lowpass filters in the literature are often based on RLC reference structures, with the use of active inductance circuits to substitute the physical inductors. However, implementations based on the Gm-C [6] approach are quite common, and filters based on the closed loop Sallen–Key [7] and Tow–Thomas [8] topologies have also been reported in the low-GHz range. Closed loop filter architectures based on non-conventional active building blocks—such as second-generation current conveyors [9] or second-generation voltage conveyors [10]—have been also exploited at lower frequencies. However, very few lowpass filter implementations above 4 GHz are reported in the literature, and none of them are based on closed-loop architectures. In [11], a tunable 5th order elliptic Gm-C lowpass filter in 170 GHz-f_T SiGe BiCMOS with a maximum bandwidth of 4.1 GHz was reported, and in [12] a 3rd order Gm-C filter with a maximum bandwidth of 10 GHz, in 65-nm CMOS was reported. Filters based on the active inductance approach have been presented in [13], that reports a 5th order 4.57-GHz lowpass filter in 180-nm CMOS, and [14] that describes a 10.5-GHz biquad in SiGe HBT technology.

On the other hand, the ever-increasing frequency performance of advanced bipolar technologies and deep submicron CMOS allows achieving huge gain-bandwidth products, thus making it possible to adopt a closed-loop approach for the design of multi-GHz filters. This allows using filter design techniques that are typically adopted at lower frequencies, both for the topology of the basic filter stage, the biquad, and for the system design of higher order filters under technology constraints (e.g., limits on the maximum quality factor that can be achieved) [15,16]. The closed-loop approach offers the advantages of increased linearity and low sensitivity to active devices variations, thanks to feedback; the filter characteristics are related to the values of passive components and/or to their ratios, and could be easily tuned, e.g., by using varactors.

In this paper, we demonstrate a 10 GHz, fully differential, biquadratic filter exploiting Sallen–Key architecture and based on a differential difference amplifier (DDA). The proposed DDA design makes use of both the bipolar and MOS transistors available in the adopted commercial 55-nm SiGe BiCMOS technology. It must be noted that this is the first work in which a closed-loop approach is used to design filters at such high frequencies, resulting in an improved linearity with a power consumption comparable to alternative approaches.

In the following sections, Section 2 describes the proposed biquad architecture and design equations, Section 3 presents the detailed design of the DDA amplifier, Section 4 deals with filter design referring to the adopted 55-nm BiCMOS process, Section 5 summarizes the results of the simulations, and, finally, Section 6 concludes this work.

2. Proposed Biquad Architecture

Figure 1 shows the proposed fully differential topology of the Sallen–Key (*SK*) lowpass biquad stage based on a differential difference amplifier (DDA). The DDA can be considered a fully differential amplifier with two differential input pairs. *SK* filters have two feedback loops: negative feedback is used to determine the low-frequency gain, and positive feedback allows us to determine the frequency response, thanks to the feedback network formed by capacitors C_1 and C_2 and the two resistors R_1 and R_2.

The transfer function of the circuit in Figure 1, assuming an ideal DDA, can be easily computed as:

$$\frac{V_{out}}{V_{in}} = \frac{\frac{G}{C_1 C_2 R_1 R_2}}{s^2 + \left(\frac{1}{R_1 C_1} + \frac{1}{R_2 C_1} + \frac{1-G}{R_2 C_2}\right)s + \frac{1}{C_1 C_2 R_1 R_2}} \tag{1a}$$

$$G = 1 + \frac{R_B}{R_A} \tag{1b}$$

Hence, the quality factor and resonance frequency of the lowpass filter are:

$$\begin{cases} f_0 = \frac{1}{2\pi \sqrt{C_1 C_2 R_1 R_2}} \\ Q = \frac{\sqrt{C_1 C_2 R_1 R_2}}{R_2 C_2 + R_1 C_2 + R_1 C_1 (1-G)} \end{cases} \tag{2}$$

From (2), it can be seen that the value of pole Q is dependent on the value of the gain G and typically the amplifier in the *SK* stage is configured to have a DC gain G larger than 1 to relax the ratio of passive components when high Q is needed. If unity gain of the filter is a requirement, then resistance R_1 can be split into two resistances forming a voltage divider (with gain $1/G <1$), thus reducing the gain from $G > 1$ to $G = 1$. However, the drawback of such a design choice is that the larger closed-loop gain limits the bandwidth of the amplifier. Therefore, the main reason to choose $G = 1$ is to maximize the closed-loop bandwidth of the DDA, which is necessary to obtain an accurate frequency response for the SK filter up to very high frequencies. Even if unity-gain feedback limits the maximum quality factor that can be achieved for a given ratio of passive components, a low-Q filter synthesis approach [15,16] can be used to synthesize high order filters with limited values for the Q of the biquad stages (typically in the range of 2 to 3). This results in an accurate frequency response and has the additional advantage of reducing the sensitivity and allowing a more robust design under PVT and mismatch variations.

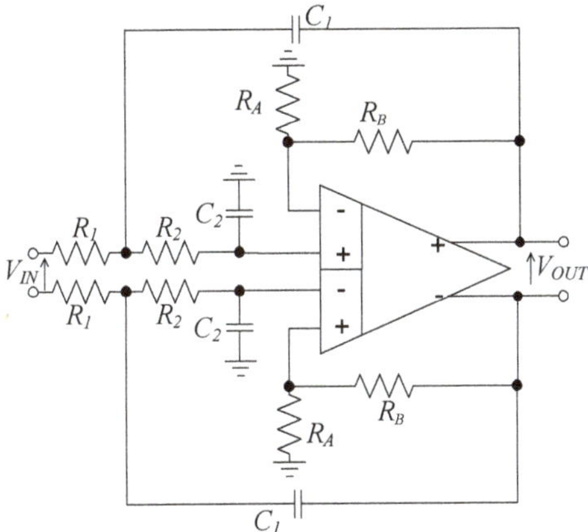

Figure 1. Architecture of the proposed lowpass biquadratic filter.

The filter architecture for $G = 1$ is reported in Figure 2, where the DDA amplifier has been configured as a fully differential closed loop voltage follower.

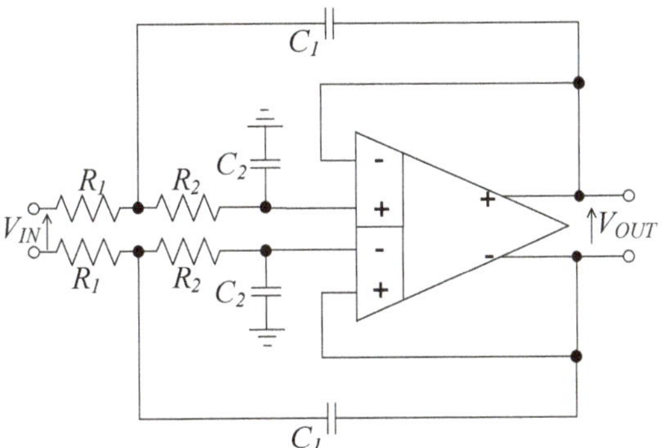

Figure 2. Architecture of the proposed lowpass filter with DDA configured as voltage follower.

The ideal frequency response of the SK stage in Figure 2 is

$$\frac{V_{out}}{V_{in}} = \frac{\frac{1}{C_1 C_2 R_1 R_2}}{s^2 + \left(\frac{1}{R_1 C_1} + \frac{1}{R_2 C_1}\right)s + \frac{1}{C_1 C_2 R_1 R_2}} \tag{3}$$

The resonance frequency and quality factor of the biquad stage can be expressed as:

$$\begin{cases} f_0 = \dfrac{1}{2\pi\sqrt{R_1 R_2 C_1 C_2}} \\ Q = \sqrt{\dfrac{C_1}{C_2}}\dfrac{\sqrt{R_1 R_2}}{R_1 + R_2} \end{cases} \tag{4}$$

The maximum Q value is achieved for $R_2 = R_1 = R$, hence this choice is optimal for achieving maximum $Q = \frac{1}{2}\sqrt{\frac{C_1}{C_2}}$.

It is important to point out that the closed loop architecture of the SK filter results in a resonance frequency and a quality factor which are ideally independent on the parameters of the active devices and depend only on the value of the passive components.

The quality factor is, under ideal conditions, stable under process and temperature variations, because it is given by the ratio between two capacitors. The resonance frequency, instead, varies with process and temperature conditions, as it is inversely proportional to variations in resistor and capacitor values. In particular, it can be easily shown that the sensitivity of the cut-off frequency of the *SK* biquad to each one of the parameters R_1, R_2, C_1 and C_2 is equal to $-1/2$, and therefore the variations of the value of integrated passive components results in a corresponding variation of the resonance frequency. However, it has to be noted that the resonance frequency can be easily tuned implementing capacitances C_1 and C_2 with varactor-diodes or tunable MOS capacitors, and the tuning voltage of variable capacitors can be used in a servo-loop or in an automating tuning loop to accurately set the value of the cut-off frequency.

$$\begin{cases} f_0 = \dfrac{1}{2\pi R\sqrt{C_1 C_2}} \\ Q = \frac{1}{2}\sqrt{\dfrac{C_1}{C_2}} \end{cases} \tag{5}$$

3. Proposed Topology for the DDA amplifier

In a single-ended implementation, the use of an opamp in non-inverting configuration allows applying both negative feedback to set the DC gain, and positive feedback to determine the frequency response. Mapping this approach to a fully differential implementation requires the use of a DDA, to make both the inverting and non-inverting differential input terminals available, whereas a standard fully differential opamp only allows the inverting configuration to be used.

Figure 3 shows the topology of the DDA used in the proposed SK biquad stage, where the different devices available in the technology have been exploited to maximize the performance: a single-stage DDA amplifier topology with output buffers (implemented as common collector stages) has been adopted to maximize the bandwidth, so to avoid the use of compensation capacitors; a common-mode feedback (CMFB) is also required, to set the output DC voltage and maximize the common-mode rejection ratio (CMRR).

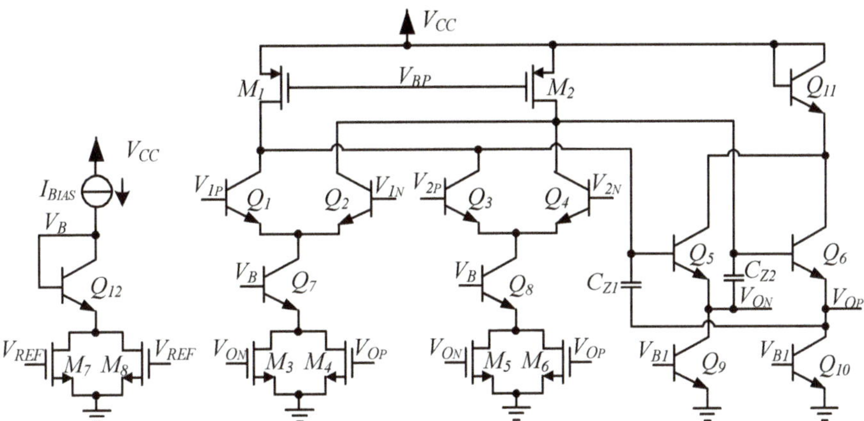

Figure 3. Proposed topology for the DDA amplifier with fully differential output.

3.1. DDA Amplifier

Referring to Figure 3, the single-stage DDA amplifier with active loads which has been used to implement the proposed biquad filter can be described as follows: high-speed npn HBT devices have been adopted for the input differential pairs (Q_1-Q_4), in order to exploit their high transconductance for a given current and outstanding frequency performance. RF PMOS transistors have been exploited as active loads to boost the DC gain without a significant frequency penalty. In the preliminary design phase, both resistive and active PMOS loads have been considered: however, when adopting a resistive load, the maximum allowed load resistance value is limited by the maximum allowable voltage drop across the resistors, and for the target supply voltage $V_{CC} = 3$ V the gain would have been too low to guarantee enough loop gain with a single amplifier stage. Furthermore, the good frequency performance of 55-nm PMOS devices results in a limited bandwidth penalty of the active load with respect to the resistive case, guaranteeing a very large gain-bandwidth product for the DDA. The output DC voltage and swing of the stage allows keeping the source-drain voltage of transistors M_1 and M_2, whose biasing voltage V_{BP} is generated by a conventional current mirror (not shown in Figure 3), below the safe limits imposed by the technology. This, however, requires a level shift to have an output DC voltage of the opamp compatible with the input DC common-mode range of the DDA.

The output stage is needed both as level shifter and to provide a very low output resistance: this is a critical issue in the design of *SK* filter stages, since the finite output resistance of the main active element of the filter reduces the maximum available quality factor. From this point of view, the implementation of the main active element of the *SK* filter as a closed loop voltage follower is advantageous, since it allows reducing the already low output resistance of the common collector stages.

In fact, remembering that the maximum Q is achieved for $R_2 = R_1 = R$ as discussed in Section 2, considering an output resistance of the closed loop DDA amplifier $R_o = \epsilon R$, and assuming a capacitance ratio $\alpha \equiv C_2/C_1$, the maximum achievable quality factor for the Sallen–Key stage can be rewritten as:

$$Q \le Q^{MAX} = \frac{\sqrt{1+\epsilon}}{\sqrt{8\epsilon}} \approx \frac{1}{\sqrt{8\epsilon}} \tag{6}$$

Hence, even the value $Q = 2$ is difficult to obtain, because it would require $R_o < R/32$.

In order to reduce R_o, high-speed HBT devices Q_5 and Q_6 have been used as common collector output buffers due to the large transconductance of bipolar devices. Q_5 and Q_6 are biased by current sources implemented with high-voltage HBT devices (Q_9 and Q_{10}) – the reference branch of the conventional current mirror that sets the bias voltage V_{B1} is not shown in Figure 3. To keep the collector-emitter voltage of the high-speed HBTs below the safe limits, a common-mode level shifter, implemented through the diode-connected transistor Q_{11}, has been exploited to reduce the collector voltage of Q_5 and Q_6. The use of a common-collector output stage is fundamental both to set the correct DC levels and to reduce the output resistance; however, the base-emitter C_π capacitance introduces a zero in the transfer function that impacts on the out-of-band behavior of the lowpass filter. To compensate this effect, cross-coupled capacitors C_{Z1} and C_{Z2} have been added, exploiting the fully differential nature of the stage to cancel out the effect of the C_π by means of positive feedback.

Neglecting r_0, C_μ, r_π in the device model and assuming a capacitive load C_L, the transfer function of the emitter follower with the cross-coupled capacitances is:

$$\frac{v_o}{v_i} = \frac{1 + s\frac{C_\pi - C_{Z1,2}}{g_m}}{1 + s\frac{C_L + C_\pi + C_{Z1,2}}{g_m}} \tag{7}$$

Hence, for $C_\pi = C_{Z1,2}$, the zero disappears, and the common collector stage only adds a pole to the transfer function, that in the limit of a large load capacitance ($C_L \gg C_\pi$) results practically unaffected by the compensation capacitor.

3.2. Common Mode Feedback Loop

A fully differential amplifier requires a common-mode feedback (CMFB) loop to set the output DC voltage and improve the common-mode rejection ratio (CMRR). In this case, a standard triode-based CMFB has been adopted: triode degeneration has been added to the current mirror that sets the current of the DDA, formed by Q_7, Q_8 and Q_{12}. The triode devices on the reference branch (M_7 and M_8) are controlled by the reference voltage (in this case, $V_{REF} = V_{CC}/2$), and the devices under the differential pairs (M_3-M_6) are controlled by the output voltages. These devices act as voltage-controlled resistors, thus adjusting the tail current of the differential pairs to match the current of the PMOS loads while setting the required output common mode voltage. Thick oxide MOS devices have been used to withstand the full swing of the output DC voltage, and their size has been optimized as a trade-off between output loading and functionality of the CMFB under PVT variations.

4. Filter Design

Unity-gain feedback and finite output conductance in the DDA, besides other non-ideal effects, limit the maximum achievable quality factor of the *SK* biquad stages. This is an important issue to cope with when adopting conventional filter synthesis techniques, because the maximum quality factor increases with the order of the filter, even in relatively low-Q designs such as Butterworth filters. Hence, non-conventional design approaches are required. In these approaches, a maximum Q value is chosen, and a filter mask is synthetized subject to this constraint. Unlike in conventional approaches, a larger number of biquad stages may be required for the same stopband attenuation. When adopting these methodologies, the maximum Q allowed for the biquad stages is typically in the range from 2 to 3 [15,16]: these values of Q are compatible with the biquad design proposed in this work.

The reduced Q value also reduces sensitivity to process and mismatch variations and to parasitic effects, as they increase for large quality factors: with lower quality factors, the frequency response is less dependent on variations in Q and f_0, resulting in a more robust design with respect to process, temperature, and mismatch variations.

The proposed filter has been designed in the SiGe BiCMOS055 technology from STMicroelectronics [17], which provides high speed, heterojunction bipolar npn transistors with values of f_T in excess of 300 GHz and an f_{MAX} in excess of 350 GHz, and RF NMOS and PMOS devices with values of f_T of about 190 GHz and 95 GHz, respectively. High voltage bipolar and MOS transistors with reduced frequency performance are also available to implement current sources and biasing circuits.

The biquadratic filter has been designed for a resonance frequency $f_0 = 6.4$ GHz and a quality factor $Q = 2$, as the basic building block of a low-Q higher order filter for anti-aliasing applications in high-speed digitizers [4,5]. A resonant frequency of 6.4 GHz is ideally equivalent to a 3-dB bandwidth (f_{3dB}) of 10 *GHz* for $Q = 2$.

Table 1 shows the device sizing and bias current of all the bipolar and MOS transistors of the circuit in Figure 3, whereas the values of filter components (referring to Figure 2) have been set as follows: $R_1 = R_2 = 100\ \Omega$, $C_1 = 380$ fF, $C_2 = 40$ fF, $C_{Z1} = C_{Z2} = 50$ fF. R_1 and R_2 have been implemented as poly-silicon resistors, and capacitances have been implemented as MIM (metal-insulator-metal) capacitors available in the RF library of the adopted technology.

Table 1. Device sizing.

Device	Size	Bias Current
Q_1, Q_2, Q_3, Q_4	$L_e/W_e^1 = 1/0.2\ \mu m$	$IC = 1$ mA
Q_5, Q_6	$L_e/W_e = 4/0.2\ \mu m$	$IC = 1$ mA
Q_7, Q_8, Q_{12}	$L_e/W_e = 5/0.2\ \mu m$	$IC = 1$ mA
Q_9, Q_{10}	$L_e/W_e = 5/0.2\ \mu m$	$IC = 1$ mA
Q_{11}	$L_e/W_e = 2 \times 4/0.2\ \mu m$	$IC = 1$ mA
M_1, M_2	$W/L = 6/0.06\ \mu m$	$ID = 1$ mA
M_3, M_4, M_5, M_6	$W/L = 20/0.25\ \mu m$	$ID = 1$ mA

$^1\ L_e/W_e$ = Emitter Length/Emitter Width

The layout of the filter is reported in Figure 4: the filter occupies an area of only $65 \times 42\ \mu m^2$.

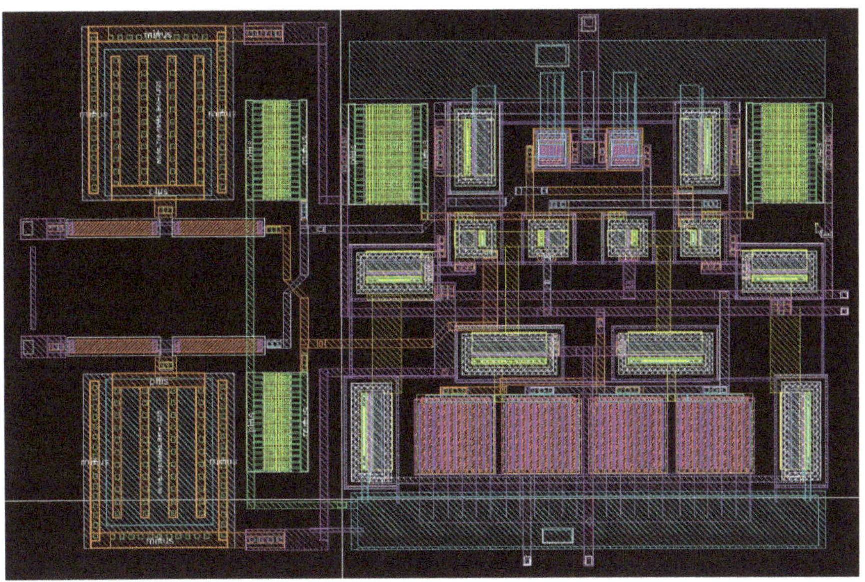

Figure 4. Layout of the proposed Biquad Filter.

5. Simulation Results

Table 2 shows simulation results for the typical process corner and temperature (T = 27 °C). The three FOMs reported in Table 2 are defined as follows:

$$FOM_1 = \frac{P_{diss}}{N_{pole}} \tag{8a}$$

$$FOM_2 = \frac{P_{diss}}{N_{pole}f_{3dB}} \tag{8b}$$

$$FOM_3 = \frac{P_{diss}}{N_{pole}f_{3dB}D_R} \tag{8c}$$

where N_{pole}, and D_R denote the number of poles (e.g., the order) and the dynamic range (computed as in [14]) of the filter respectively.

Table 2. Typical 27 °C simulations.

Name	Value	Unit	Notes
P_{diss}	18	mW	Power Dissipation
V_{cc}	3	V	Supply Voltage
f_0	6.45	GHz	Resonant Frequency
f_{3dB}	9.55	GHz	3 dB Bandwidth
A_{DC}	−0.5	dB	DC-Gain
A_{PK}	5.8	dB	Peak Gain
Q	2.06	-	Quality Factor
SFDR	64	dB	@0.8Vpp differential,1 GHz
v_{onoise}	1.36	mVrms	Output Noise integrated between 1 Hz and 10 GHz
SNR	46.4	[dB]	@0.8Vpp differential
D_R [1]	49.3	dB	Dynamic Range
FOM1	9	mW	
FOM2	0.94	pW/Hz	
FOM3	0.0032	mW/GHz	

[1] D_R has been computed as in [14].

Figure 5 shows the frequency response of the filter. The Discrete Fourier Transform (DFT) of the differential output voltage with an input tone at 1 GHz, 0.8Vpp differential is reported in Figure 6.

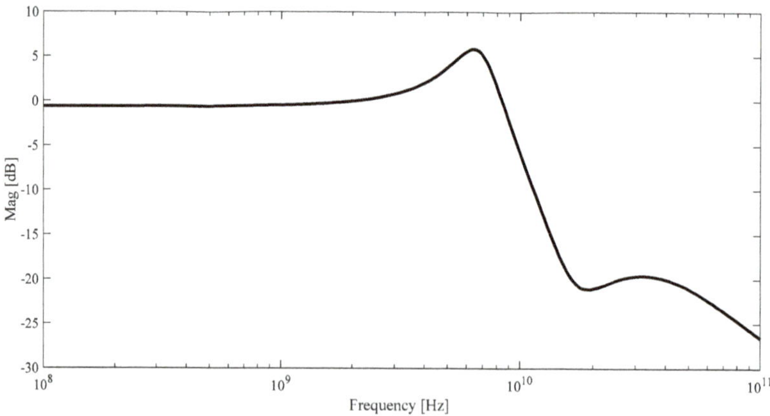

Figure 5. Frequency Response of the proposed Biquadratic Filter.

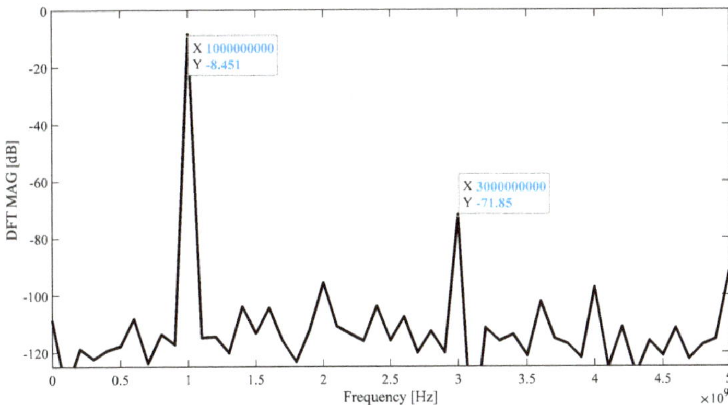

Figure 6. Discrete Fourier Transform (DFT) of the differential output voltage with an input tone at 1GHz, 0.8Vpp differential.

The DFT result reported in Figure 6 demonstrates that linearity is very good, with 64 dB SFDR in the relatively large input signal condition of 800 mVpp. The dynamic range is dominated by noise, as SFDR is much better than SNR (see Table 2).

Table 3 shows the results of the simulations accounting for temperature variations, and Table 4 shows those accounting for supply voltage variations. These results confirm the robustness of the filter to both temperature and supply voltage variations.

Table 3. Simulations vs. temperature.

Name	Minimum	Maximum	Unit
T	−30	120	°C
P_{diss}	18.4	19.56	mW
f_0	6.91	5.65	GHz
f_{3dB}	10.2	8.2	GHz
A_{pk}	5.51	5.3	dB
A_{dc}	−0.53	−0.58	dB
Q	2	1.9	-
SFDR	61	73	dB
v_{onoise}	1.12	1.51	mV
D_R [1]	49.4	51.6	dB

[1] D_R has been computed as in [14].

Table 4. Simulations vs. supply voltage.

Name	Value	Value	Unit
V_{cc}	2.85	3.15	V
P_{diss}	17.1	19.9	mW
f_0	6.1	6.4	GHz
f_{3dB}	8.9	9.6	GHz
A_{pk}	4.89	5.72	dB
A_{dc}	−0.7	−0.56	dB
Q	1.9	2	-
SFDR	48	56	dB
v_{onoise}	1.26	1.26	mV
D_R [1]	44.3	47	dB

[1] D_R has been computed as in [14].

Monte Carlo simulations, using the accurate statistical models for HBT, MOS and passive devices available in the design kit of the BiCMOS055 technology, have been carried out in the Cadence Virtuoso ADE XL environment. Figure 7 shows the frequency response of the biquad filter for 100 Monte Carlo iterations, whereas Figures 8–11 show the histograms of DC gain, quality factor, resonant frequency and 3 dB bandwidth of the filter respectively. A summary of the mean values (μ) and standard deviations (σ) for these quantities is reported in Table 5.

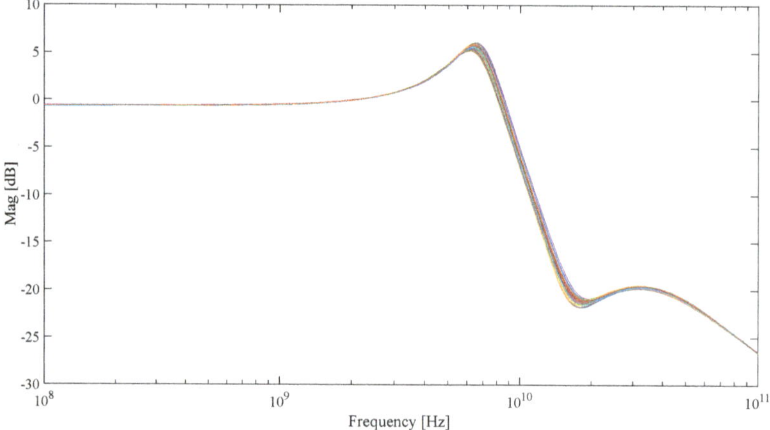

Figure 7. Frequency Response for 100 Monte Carlo iterations.

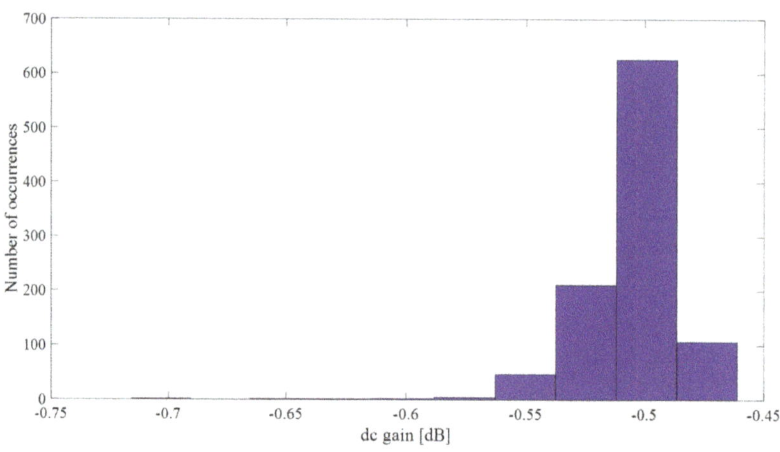

Figure 8. Histogram of DC Gain A_{dc} for 1000 Monte Carlo iterations.

Table 5. Summary of Monte Carlo simulations (1000 iterations).

Name	μ	σ	Unit
A_{dc}	−0.505	0.018	dB
Q	2.071	0.037	-
f_0	6.471	0.100	GHz
f_{3dB}	9.588	0.126	dB

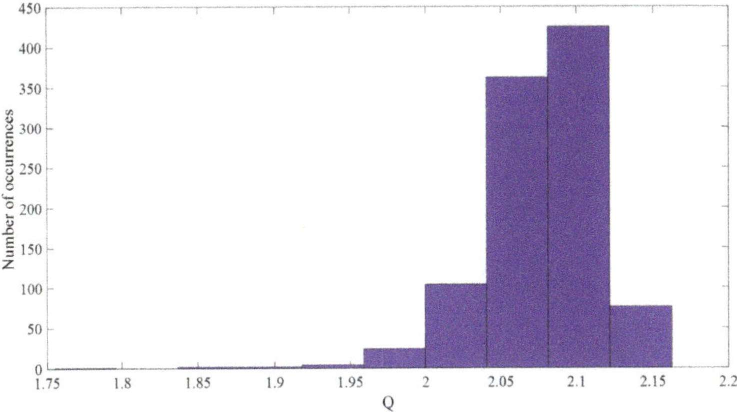

Figure 9. Histogram of quality factor Q for 1000 Monte Carlo iterations.

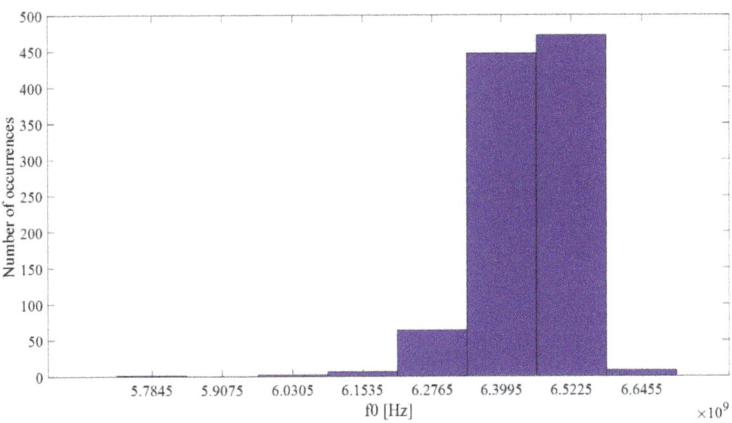

Figure 10. Histogram of resonant frequency f_0 for 1000 Monte Carlo iterations.

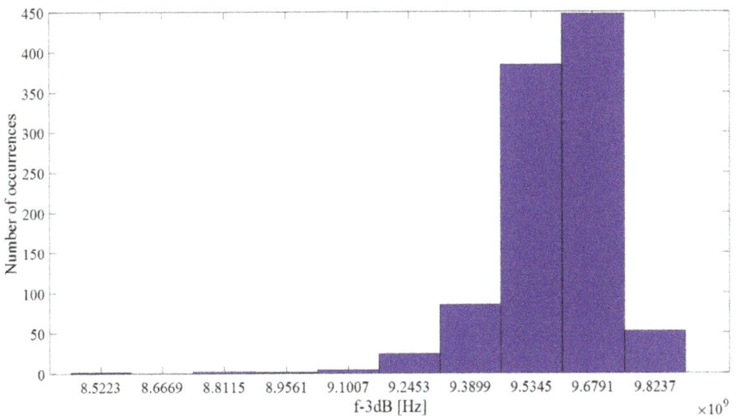

Figure 11. Histogram of 3 dB bandwidth f_{3dB} for 1000 Monte Carlo iterations.

All these results allow us to assess the robustness of the filter under Monte Carlo simulations: the DC gain is almost constant, while the quality factor and resonance frequency of the complex poles vary limitedly.

A comparison with recent technical papers reporting inductorless active lowpass filters with cut-off beyond 1.5 GHz and the proposed filter is reported in Table 6. The proposed filter outperforms all the other designs in terms on SFDR, DR and FOM3, confirming the outstanding linearity performance of our unitary-gain, closed-loop approach.

Table 6. Comparison against the state of the art.

	This Work	[14]	[8]	[18]	[12]	[19]	[11]	[13]
Year	2020	2018	2017	2014	2012	2010	2006	2002
Technology	BiCMOS 55 nm	BiCMOS 55 nm	CMOS 40 nm	CMOS 28 nm	CMOS 65 nm	BiCMOS 180 nm	BiCMOS 250 nm	CMOS 180 nm
Filter Type	*SK*	*AL*	*TT*	*GMC*	*GMC*	*GMC*	*GMC*	*AL*
Supply Voltage [V]	3	3	1.1	1.1	1.4	3.3	3.5	1.8
P_{diss} [mW]	18	13.7	17.6	30	140	300	99.8	1.6
f_{3dB} [GHz]	9.55	10.5	1.6	3.3	10	3.2	4.1	4.6
SFDR [dB])	64	−51	−46	−40	−45	40	-	−57
v_{onoise} [mVrms]	1.36	1.1	1.5	-	0.5	-	-	-
SNR [dB]	46.4	47.2	43.2	39	45	-	-	-
D_R^1 [dB]	49.3	45.5	41.1	37	42	-	20.1	-
Npole	2	2	5	5	3	6	5	5
FOM1 [mW]	9	6.85	3.52	6	46.6	50	19,9	0.32
FOM2 [pW/Hz]	0.94	0.65	2.2	1.81	4.7	15.6	4,8	0.07
FOM3[mW/GHz]	0.0032	0.0035	0.019	0.025	0.037	-	0.47	-
Area (mm^2)	0.0027	0.0056	0.12	0.091	0.01	0.17	0.82	-
Measured	No	No	Yes	Yes	Yes	No	Yes	Yes

[1] D_R has been computed as in [14].

It has to be noted that most papers in the literature are based on the Gm-C approach [11,12,18,19], in which the open-loop input stage often limits linearity, whereas the closed loop approach proposed in this work exploits feedback to reduce nonlinear distortions. Focusing on reference [8], it is based on a closed loop Tow–Thomas (TT) architecture which shows a limited dynamic range owing to the lower bandwidth of CMOS amplifiers, which makes feedback less effective. Reference [14] exploits positive feedback to synthetize an active inductor (AL), and resistive degeneration of the differential pair to improve linearity. Having similar performance and using the same technology, this solution shows lower power consumption but also lower dynamic range; hence, it performs better in the first two FOMs, but worse in the third-one, owing both to worse SNR and SFDR performance. Reference [13] reports a 5th-order filter based on an active circuit which provides a transfer function with two poles and two zeros with only three active devices: this provides for very good power efficiency, as shown by excellent FOM1 and FOM2 performance, but no detailed noise information is provided, so FOM3 cannot be computed. As it reports 100nV/$\sqrt{Hz}$ noise for 4.57 GHz of bandwidth, estimated noise is 6.8 mV, which would provide a very low SNR of 25.5 dB. Hence, the dynamic range of such a design is severely limited.

6. Conclusions

A fully differential Sallen–Key filter with a 3-dB bandwidth of about 10 GHz and based on a DDA amplifier which exploits bipolar and MOS devices of the commercial BiCMOS055 technology from STMicroelectronics has been proposed in this work. Post-layout simulations have shown that the proposed filter outperforms all recently published inductorless active lowpass filters with cut-off frequencies beyond 1.5 GHz in terms of *SFDR*, *DR*, and *FOM3*. Parametric simulations accounting for temperature and supply voltage variations as well as Monte Carlo simulations have confirmed the robustness of the filter to temperature, supply voltage and mismatch variations. The power efficiency of the proposed filter is good and the area footprint is very low. Based on the comparison against the state of the art reported in Table 6, we have demonstrated that the proposed architecture—which exploits a fully differential DDA-based voltage follower as active element—allows the implementation of closed loop biquad filters with a 3-dB bandwidth up to about 10 GHz while guaranteeing better linearity performance with respect to filter architectures based on the Gm-C or the active inductor approach. Furthermore, we can state that it is possible to use the proposed fully differential Sallen–Key closed-loop topology for filters with bandwidths around 10 GHz, exploiting NPN devices with an f_T in the order of 300 GHz and PMOS with an f_T in the order of 100 GHz as active loads. Unity gain feedback across the DDA ensures the highest closed-loop bandwidth, though limiting the achievable maximum quality factor. The resulting filter is also compact, and shows the lowest area occupation among the existing literature.

Author Contributions: Conceptualization, G.S. and A.T.; methodology, F.C. and P.M.; software, P.T.; validation, F.C., P.M. and G.S.; formal analysis, P.M.; investigation, F.C.; resources, A.T.; writing—original draft preparation, G.S.; writing—review and editing, P.T., F.C. and P.M.; funding acquisition, A.T. All authors have read and agreed to the published version of the manuscript.

Funding: This work is supported by European ECSEL-JU/EU-H2020 under grant no. 737454.

Conflicts of Interest: The authors declare no conflict of interest.

References

1. Kim, S.-N.; Kim, W.-C.; Seo, M.-J.; Ryu, S.-T. A 65-nm CMOS 6-Bit 20 GS/s Time-Interleaved DAC With Full-Binary Sub-DACs. *IEEE Trans. Circuits Syst. II Express Briefs* **2018**, *65*, 1154–1158. [CrossRef]
2. Zandieh, A.; Schvan, P.; Voinigescu, S.P. Design of a 55-nm SiGe BiCMOS 5-bit Time-Interleaved Flash ADC for 64-Gbd 16-QAM Fiberoptics Applications. *IEEE J. Solid-State Circuits* **2019**, *54*, 2375–2387. [CrossRef]
3. Sun, K.; Wang, G.; Zhang, Q.; Elahmadi, S.; Gui, P. A 56-GS/s 8-bit Time-Interleaved ADC With ENOB and BW Enhancement Techniques in 28-nm CMOS. *IEEE J. Solid-State Circuits* **2018**, *54*, 821–833. [CrossRef]
4. Monsurrò, P.; Trifiletti, A.; Angrisani, L.; D'Arco, M. Streamline calibration modelling for a comprehensive design of ATI-based digitizers. *Measurement* **2018**, *125*, 386–393. [CrossRef]
5. Monsurrò, P.; Trifiletti, A.; Angrisani, L.; D'Arco, M. Two novel architectures for 4-channel mixing/filtering/processing digitizers. *Measurement* **2019**, *142*, 138–147. [CrossRef]
6. Chen, Y.; Mak, P.-I.; Wang, Y.; Zhang, L.; Qian, H. 0.013 mm2, kHz-to-GHz-bandwidth, third-order all-pole lowpass filter with 0.52-to-1.11 pW/pole/Hz efficiency. *Electron. Lett.* **2013**, *49*, 1340–1342. [CrossRef]
7. Wambacq, P.; Giannini, V.; Scheir, K.; Van Thillo, W.; Rolain, Y. A fifth-order 880MHz/1.76GHz active lowpass filter for 60GHz communications in 40nm digital CMOS. In Proceedings of the ESSCIRC 2010—36th European Solid State Circuits Conference, Seville, Spain, 14–19 September 2010; pp. 350–353.
8. Wu, C.-D.; Hsieh, J.-Y.; Wu, C.-H.; Cheng, Y.-S.; Wu, C.-C.; Lu, S.-S. An 1.1 v 0.1–1.6 GHz tunable-bandwidth elliptic filter with 6 dB linearity improvement by precise zero location control in 40 nm CMOS technology for 5G applications. In Proceedings of the 2017 IEEE International Symposium on Circuits and Systems (ISCAS), Baltimore, MD, USA, 28–31 May 2017.
9. Salem, S.B.; Fakhfakh, M.; Masmoudi, D.S.; Loulou, M.; Loumeau, P.; Masmoudi, N. A high performances CMOS CCII and high frequency applications. *Analog. Integr. Circuits Signal Process.* **2006**, *49*, 71–78. [CrossRef]
10. Safari, L.; Barile, G.; Ferri, G.; Stornelli, V. High performance voltage output filter realizations using second generation voltage conveyor. *Int. J. RF Microw. Comput. Eng.* **2018**, *28*, e21534. [CrossRef]
11. Lu, Y.; Krithivasan, R.; Kuo, W.-M.L.; Li, X.; Cressler, J.D.; Gustat, H.; Heinemann, B. A 70 MHz–4.1 GHz 5th-order elliptic gm-C low-pass filter in complementary SiGe technology. In Proceedings of the 2006 Bipolar/BiCMOS Circuits and Technology Meeting, Maastricht, The Netherlands, 8–10 October 2006.

12. Houfaf, F.; Egot, M.; Kaiser, A.; Cathelin, A.; Nauta, B. A 65nm CMOS 1-to-10GHz tunable continuous-time low-pass filter for high-data-rate communications. In Proceedings of the 2012 IEEE International Solid-State Circuits Conference, San Francisco, CA, USA, 19–23 February 2012; pp. 362–363.

13. Xiao, H.; Schaumann, R. Very-high-frequency lowpass filter based on a CMOS active inductor. In Proceedings of the 2002 IEEE International Symposium on Circuits and Systems, Phoenix-Scottsdale, AZ, USA, 26–29 May 2002.

14. Centurelli, F.; Monsurrò, P.; Trifiletti, A. A 10 GHz inductorless active SiGe HBT lowpass filter. *Int. J. RF Microw. Comput. Eng.* **2018**, *28*, e21567. [CrossRef]

15. Monsurrò, P.; Pennisi, S.; Scotti, G.; Trifiletti, A. Design strategy for biquad-based continuous-time low-pass filters. In Proceedings of the 20th European Conference on Circuit Theory and Design, Linkoping, Sweden, 29 August 2011; pp. 385–388.

16. Monsurrò, P.; Pennisi, S.; Scotti, G.; Trifiletti, A. High-tuning-range CMOS band-pass IF filter based on a low-Qcascaded biquad optimization technique. *Int. J. Circuit Theory Appl.* **2014**, *43*, 1615–1636. [CrossRef]

17. Chevalier, P.; Avenier, G.; Canderle, E.; Montagné, A.; Ribes, G.; Vu, V.T. Nanoscale SiGe BiCMOS technologies: From 55 nm reality to 14 nm opportunities and challenges. In Proceedings of the 2015 IEEE Bipolar/BiCMOS Circuits and Technology Meeting – BCTM, Boston, MA, USA, 26–28 October 2015.

18. Sabatino, N.; Minoia, G.; Roche, M.; Baldi, D.; Temporiti, E.; Mazzanti, A. A 5th Order gm-C Low-Pass Filter with ±3% Cut-Off Frequency Accuracy and 220 MHz to 3.3 GHz Tuning-Range in 28nm LP CMOS. In Proceedings of the IEEE European Solid State Circuits Conference, ESSCIRC 2014, Venice Lido, Italy, 22–26 September 2014; pp. 351–354.

19. Baranauskas, D.; Zelenin, D.; Bussmann, M.; Elahmadi, S.; Edwards, J.K.; Gill, C.A. A 1.6-3.2-GHz sixth-order +13.1-dBm OIP3 linear phase gm-C filter for fiber-optic EDC receivers. *IEEE Trans Microw Theory Tech.* **2010**, *58*, 1314–1322. [CrossRef]

Article

Compact Ultra-Wideband Bandpass Filters Achieved by Using a Stub-Loaded Stepped Impedance Resonator

Min-Hang Weng [1], Fu-Zhong Zheng [2], Hong-Zheng Lai [2] and Shih-Kun Liu [2,3,*]

[1] School of Information Engineering, Putian University, Putian 351100, China; hcwweng@gmail.com
[2] Institute of Photonics Engineering, National Kaohsiung University of Science and Technology, Kaohsiung 80778, Taiwan; 1098319123@nkust.edu.tw (F.-Z.Z.); f107155102@nkust.edu.tw (H.-Z.L.)
[3] Doctoral Degree Program in Biomedical Engineering, Kaohsiung Medical University, Kaohsiung 80708, Taiwan
* Correspondence: skliu@nkust.edu.tw

Received: 3 January 2020; Accepted: 19 January 2020; Published: 22 January 2020

Abstract: In this paper, we develop a bandpass filter using a stub-loaded stepped impedance resonator (SLSIR) and calculate the even and odd resonant modes of this type of resonator using the input impedance/admittance analysis. In this study, two impedance ratios and two length ratios are operated as the design parameters for controlling the resonant modes of the SLSIR. Several resonant mode variation curves operating three resonant modes with different impedance ratios and two length ratios are developed. By tuning the desired impedance ratios and length ratios of the SLSIRs, compact ultra-wideband (UWB) bandpass filters (BPFs) can be achieved. Two examples of the UWB BPFs are designed in this study. The first example is UWB filter with a wide stopband and the second one is dual UWB BPF, namely, with UWB performance and a notch band. The first filter is designed for a UWB response from 3.1 to 5.26 GHz having a stopband from 5.3 to 11 GHz, with an attenuation level better than 18 dB. The second filter example is a dual UWB BPF with the frequency range from 3.1 to 5 GHz and 6 to 10.1 GHz using two sets of the proposed SLSIR. The measured results have insertion loss of less than 1 dB, and return loss greater than 10 dB. Furthermore, the coupling structures and open stub of the SLSIR also provide several transmission zeros at the skirt of the passbands for improving the passband selectivity.

Keywords: ultra-wideband; bandpass filter; stub-loaded; stepped impedance resonator

1. Introduction

The Federal Communication Commission (FCC) proposed ultra-wideband (UWB) to solve the problem of data transmission. This frequency ranges from 3.1 to 10.1 GHz. Typically, the range of operational frequency of ultra-wideband is divided into two sections: one section from 3.1 to 5 GHz, the other is from 6 to 10.1 GHz. A notch band appears at 5 to 6 GHz in the UWB from 3.1 to 10.1 GHz to avoid interfering with the signal of the popular wireless local area network (WLAN) [1].

A bandpass filter (BPF) is an essential part of the front end of wireless communications. Thus, UWB BPF is increasingly popular and has been extensively developed [2–24]. In [2,3], dual-mode ring resonators were used to obtain the UWB BPFs with controllable bandwidth adjusted by the feeding position. In [4], a mode-excited resonator was used to develop a UWB BPF with an extremely broad stopband. In [5], cross-shaped resonator was used for a UWB bandpass filter having a sharp skirt and notched band. In [6,7], a UWB BPF with a wide stopband was proposed by using defected ground structure (DGS). In [8,9], parallel coupled lines were developed to have wideband filter responses with high selectivity. In [10], transmission lines implemented on metamaterial-inspired co-planar waveguide (CPW) balanced cells were utilized to obtain wideband performance. Moreover, in [11], multiple-mode split-ring resonators in a rectangular waveguide cavity were realized to achieve a wideband BPF design. However, most of above designs suffer from decreased design freedoms to tune the bandwidth and band selectivity.

Recently, multiple-mode resonator (MMR) have become commonly used as basic building blocks for wideband BPFs. Especially, stepped-impedance resonator (SIR) and stub-loaded resonator (SLR) are two most-general MMRs. The standard SIR is symmetrical structure with non-continuous impedance shaped like a dumbbell. SIR can adjust the harmonics by changing the length ratio and impedance ratio to achieve multiband or wideband performance. In [12], asymmetric SIR was used to achieve a compact UWB BPF having good band selectivity and wide stopband performance. In [13], an embedded SIR was used to achieve wideband response with a notch band. SLR structures are typically categorized into two types: open ended and short ended. In most SLR structures, the fundamental resonant frequency is determined by the main resonator and the other higher frequencies can be determined by adjusting the stub. The combination of the conventional resonator with the stub load able to have more design freedoms. In [14–16], stub-loaded multiple-mode resonators were employed to obtain UWB BPF with improved in-band performance. Moreover, in [17,18], stepped-impedance stub-loaded resonator (SISLR) was presented to have more design parameters to control the wideband responses such as bandwidth and selectivity. In [18,19], a quadruple mode ring resonator and penta-mode resonator was presented to achieve sharp-rejection broadband BPF. In [20–24], various types of stub-loaded resonators such as C-shaped and E-shaped resonators, and multi-layered substrate with two stub-loaded resonators were developed to design UWB BPF with notched bands to have high band selectivity. In previous studies, SIR combined with an open stub load was implemented to obtain dual- [25] and tri-band BPFs [26]. However, the analysis of the SIR combined with an open stub load was not described in detail.

In this paper, a stub-loaded stepped impedance resonator (SLSIR) is developed, as shown in Figure 1. The developed SLSIR was constructed from a conventional two-step SIR added with an open circuited stub at the symmetry line of the SIR. It is well known the SIR will shift the higher order modes far away or near the fundamental mode as the impedance ratio becomes larger than 1 or less than 1 [27]. The input impedance/admittance analysis is employed to obtain the even and odd resonant modes of this type of resonator in detail. For controlling the resonant modes of the SLSIR, two impedance ratios and two length ratios can be used as the design parameters. By using the impedance/admittance analysis, several variation curves with all possible resonant modes are obtained. Two types of compact UWB BPFs are designed by applying only the SLSIR with tuning of the desired impedance ratios and length ratios. In the first filter example, a UWB filter with a wide stopband filter is designed. The first odd mode and even mode constitute a UWB response from 3 to 5.1 GHz, and the other modes are shifted to far away from the passband to form a wide stopband with an attenuation better than -18 dB from 5.9 to 12 GHz. The passband edge has three transmission zeros, making high attenuation and isolation. In the second filter example, a dual ultra-wideband BPF is designed. The first odd mode and even mode in the proposed resonator constitute a passband of the ultra-wideband. Therefore, the result of the passband can be expected. The passband edge of the filter has five transmission zeros, also providing a good attenuation and isolation. This paper is organized as follows: The introduction section describes the background, motivation and novelty of this study. Several design methods of the UWB BPF are reviewed; the second section analyzes the SLSIR and develops several resonant mode variation curves of SLSIR; the third section describes the design of UWB filter with a wide stopband; the forth section describes the design of dual UWB filter (namely, a UWB with a notch band). Two filter examples were fabricated, and the measured results are found to be in good agreement with the simulated results. In conclusion, the benefits and new discovery of this design are made and addressed.

2. Analysis of the SLSIR

Figure 1 shows the structure of the SLSIR which first appeared in [10]. However, the structure is only used to design a dual-band filter. The reported design is the simulated result without accompanying theoretical analysis.

In this paper, theoretical analysis for the structure is orderly executed by the transmission line theory. In Figure 1a, the Z_1, Z_2 and Z_S denote the particular characteristic impedance of the stepped impedance resonator and the open stub, and θ_1, θ_2 and θ_S represent the electronic length of the stepped impedance resonator and the open stub, respectively.

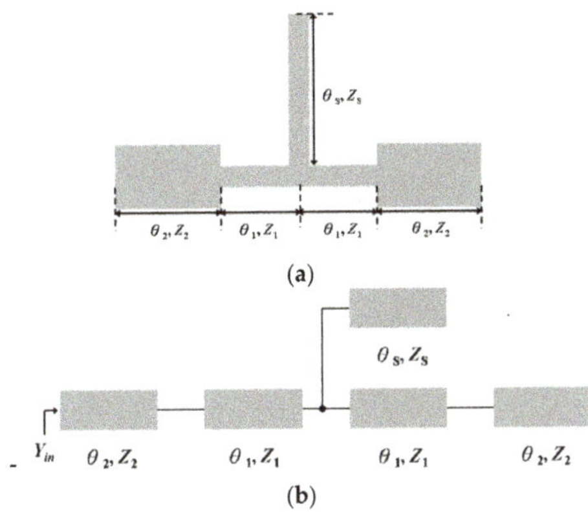

Figure 1. (a) The structure and (b) the equivalent block of the proposed stub-loaded stepped impedance resonator.

The structure of the SLSIR is mirror symmetry, and it has six variables to be controlled. Figure 1b is the equivalent block of the SLSIR. The input admittance (Y_{in}) can be calculated by transmission line theory as the following [28]:

$$Y_{in} = \frac{1}{Z_2} \frac{Z_1(K_1 - \tan\theta_1 \tan\theta_2) + jZ_L(K_1 \tan\theta_1 + \tan\theta_2)}{Z_L(1 - K_1 \tan\theta_1 \tan\theta_2) + jZ_1(\tan\theta_1 + K_1 \tan\theta_2)} \tag{1}$$

where $Z_L = \dfrac{jZ_1 K_2(\tan\theta_1 \tan\theta_2 - K_1)}{K_2(\tan\theta_2 + K_1 \tan\theta_1) - \tan\theta_s(\tan\theta_1 \tan\theta_2 - K_1)}$; $K_1 = \dfrac{Z_2}{Z_1}$; $K_2 = \dfrac{Z_S}{Z_1}$

K_1 and K_2 are two defined impedance ratios, regarding to the SIR and stub, respectively. The resonant modes of the formula are derived by setting Y_{in} equal to zero, as following [29]:

$$(K_1 - \tan\theta_1 \tan\theta_2) = 0 \tag{2}$$

and

$$[2K_2(K_1 \tan\theta_1 + \tan\theta_2) + \tan\theta_s(K_1 - \tan\theta_1 \tan\theta_2)] = 0 \tag{3}$$

It is shown Equations (2) and (3) are related to the odd- and even-mode resonances of the SLSIR, respectively. Since there are three electronic lengths θ_1, θ_2 and θ_S in the formula, two other length ratios can be defined to gain more design freedom. The first length ratio (R_1) of the SIR is defined as $R_1 = 2\theta_2/2(\theta_1+\theta_2) = 2\theta_2/\theta_T$ and the second length ratio (R_2) of the stub with SIR is defined as $R_2 = 2\theta_S/\theta_T$, where θ_T is the total length of the SIR section defined as $2(\theta_1+\theta_2)$, and θ_s is the length of the stub. Thus, R_1 and R_2 can also be varied to tune the higher order resonant modes in a wide frequency range. Thus, by setting the length ratio (R_1) and the length ratio (R_2) into Equations (2) and (3), resonant modes of the proposed can be controlled by four varied parameters, (K_1 and K_2) and (R_1 and R_2).

Figure 2 shows the length ratio R_1 with resonant electric length θ_T using MATLAB at various length ratios, $R_2 = 0.1, 0.3, 0.5, 0.7$ and 0.9. Z_s is set to be equal to Z_1, i.e., $K_2 = 1$ to reduce the plot complexity.

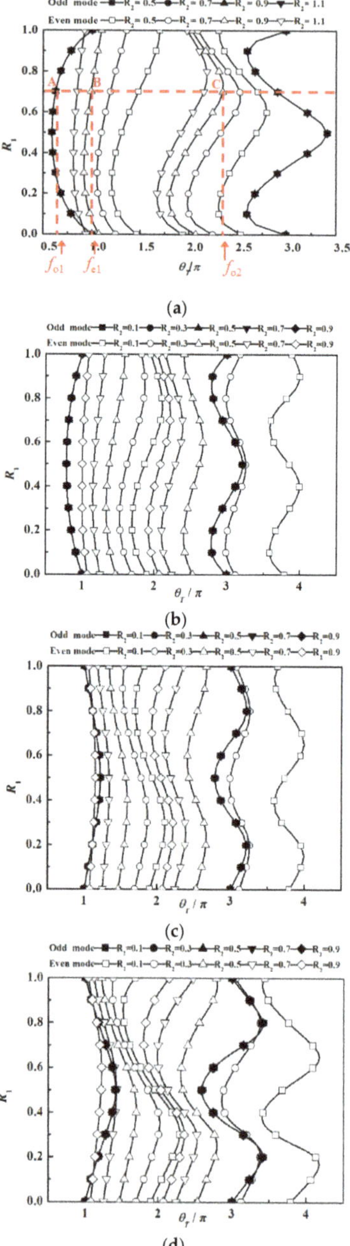

Figure 2. Length ratio R_1 with resonant electric length θ_T using the MATLAB tool at various length ratios, $R_2 = 0.1, 0.3, 0.5, 0.7$ and 0.9 based on (**a**) $K_1 = 0.25$ and $K_2 = 1$, (**b**) $K_1 = 0.5$ and $K_2 = 1$, (**c**) $K_1 = 2$ and $K_2 = 1$ and (**d**) $K_1 = 4$ and $K_2 = 1$.

As shown in Figure 2, multi-resonant modes are varied as four parameters. It is clearly found that the length ratio R_2 would not affect the resonances of the odd modes since the stub is ignored under the odd mode excitation; and the length ratio R_2 affects the resonances of the even mode obviously. The fundamental odd mode shifts to lower frequency for $K_1 < 1$ (shown in Figure 2a,b) and to higher frequency for $K_1 > 1$ (shown in Figure 2c,d), as similar to the conventional SIR. Especially, in the condition of $K_1 > 1$, at the lowest resonant frequency the even and odd modes of the proposed SLSIR join together, indicating different behavior from the resonant modes of the typical SIR. Therefore, by using these resonant curves, the structure can be designed for the different filters, for example dual-band BPF, tri-band BPF, UWB BPF and wide stopband BPF, by controlling the multi-resonant modes.

In this study, two types of UWB BPFs are designed by applying the SLSIR by tuning the desired impedance ratios and length ratios.

3. UWB BPF with Wide Stopband

In the first example, a UWB BPF with a wide stopband filter is presented. Figure 3 shows the structure of the designed filter. It mainly comprises a SLSIR with input/output coupled lines. The required passband of this BPF is set from 3.1 to 5.2 GHz, where the stopband is designed from 5 to >10 GHz.

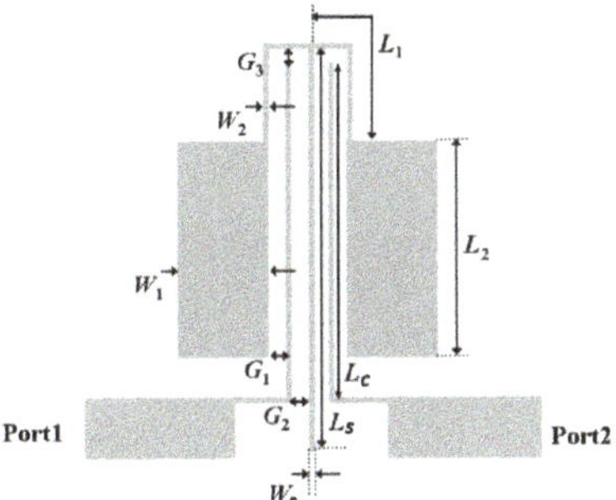

Figure 3. Structure of the ultra-wideband bandpass filter with a wide stopband.

3.1. Design Procedure

For achieving the required specification, the first odd mode and even mode are selected to constitute a UWB response from 3 to 5.2 GHz, and the other modes shall be shifted to far away from the passband to form a wide stopband. In this study, $K_1 = 0.25$ and $K_2 = 1$, as the mapping in Figure 2a is used. In this design, the first even mode (f_{e1}) and odd mode (f_{o1}), as shown in spot A and spot B in Figure 2a respectively, are set to form the UWB response. The desired resonant frequencies of the SLSIR are set as: $f_{o1} = 3.1$ GHz and $f_{e1} = 4.8$ GHz. For a wide stopband, f_{o2} is set to be greater than 10 GHz, as shown in spot C in Figure 2a. Namely, the passband frequency ratios f_{e1}/f_{o1} is 1.55 and f_{o2}/f_{o1} greater than 3.2 is required. There would be many solutions to satisfy the required conditions, thus many groups (R_1, R_1) can be selected in various groups (K_1, K_2). As mapping in Figure 2a, the corresponding length ratio R_1 is found to be 0.7 and length ratio R_2 is found to slightly more than 0.9, where the passband frequency ratios of $f_{e1}/f_{o1} = 1.58$ and f_{o2}/f_{o1} is greater than 4.64.

Both of the high impedance Z_1 and Z_S are set to be 152 Ω, and thus the low impedance Z_2 is set to be 38 Ω. According to Figure 2a, the electrical length θ_1, θ_2, and θ_S are decided as 22°, 50°, and 113°, respectively. Thus, the structure parameters of the SLSIR are obtained as: $L_1 = 3.35$ mm, $L_2 = 7.1$ mm, $L_S = 13.7$ mm, $W_1 = 3.38$ mm, $W_2 = 0.2$ mm, $W_s = 0.2$ mm. In order to miniaturize the filter size, the proposed SLSIR is folded and the input coupled line and output coupled line with high impedance are inserted into the SLSIR, as shown in Figure 3, to have an enough coupling. Therefore, the first even modes (f_{e1}) and odd modes (f_{o1}) can be coupled together to form the UWB. Figure 4 displays the simulated results of the UWB filter with a wide stopband using a full wave electromagnetic (EM) simulation [30]. When setting $G_1 = 0.2$ mm, $G_2 = 0.15$ mm and $G_3 = 0.15$ mm, the simulated insertion loss $|S_{21}|$ and return loss $|S_{11}|$ are around 0.56 dB and 14.9 dB respectively. Further, the 3-dB fractional bandwidth (FBW) is about 53.0%, and the wide stopband is from 5.9 to 11 GHz with an attenuation larger than −18 dB.

The proposed filter has the advantage of the appearance of the transmission zeros, which are produced by the open stub (L_s) and the coupling length (L_c) of the I/O lines. The length of the open stub is represented as L_s, and the length of the coupling surface is represented as L_c. The transmission zeros of the filter generated by the L_s is expressed as [29]

$$f_Z = (2n-1)f_0 \frac{\lambda_g}{4 \times L_s} (n = 1, 2, 3, \cdots) \tag{4}$$

The transmission zeros of the filter generated by the L_c is expressed as [29]

$$f_Z = (n-1)f_0 \frac{\lambda_g}{4 \times L_c} (n = 1, 2, 3, \cdots) \tag{5}$$

where f_0 is the center frequency and λ_g is the guided wavelength corresponding to the center frequency. With $L_s = 13.7$ mm and $L_c = 11.9$ mm, the expected transmission zeros are calculated to be 2.48 GHz (TZ_1) by using Equation (4), and the transmission zeros are determined to be 2.90 GHz (TZ_2) and 5.95 GHz by (TZ_3) using Equation (5). The passband edge has three transmission zeros, making a good band selectivity, as shown in Figure 4.

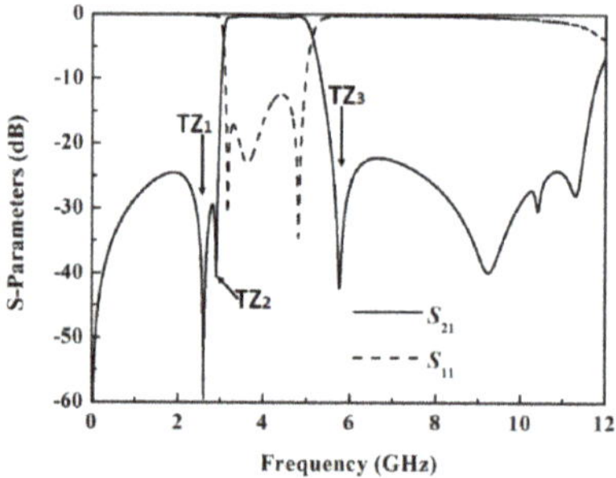

Figure 4. Simulated results of the UWB filter with a wide stopband using EM simulation. $L_1 = 3.35$ mm, $L_2 = 7.1$ mm, $L_s = 13.7$ mm, $L_c = 11.9$ mm, $W_1 = 3.38$ mm, $W_2 = 0.2$ mm, $W_s = 0.2$ mm, $G_1 = 0.2$ mm, $G_2 = 0.15$ mm, $G_3 = 0.15$ mm.

3.2. Measured Results

The designed filter is realized on 0.787 mm thick substrate (RT/Duroid 5880) having a dielectric constant and a loss tangent of 2.2 and 0.0009, respectively. The filter is optimized using a full wave EM simulation [30] and the dimension parameters are: $L_1 = 3.35$ mm, $L_2 = 7.1$ mm, $L_s = 13.7$ mm, $W_1 = 3.38$ mm, $W_2 = 0.2$ mm, $W_s = 0.2$ mm, $G_1 = 0.2$ mm, $G_2 = 0.15$ mm, $G_3 = 0.15$ mm.

Figure 5a shows a photograph of the fabricated BPF. The overall size of this BPF is about 10 mm × 20 mm, around 0.15 λ_g × 0.25 λ_g, where λ_g is the guided wavelength at the center frequency. Figure 5b displays the simulated and measured results of the designed UWB BPF with a wide stopband. For the measurement, an HP 8510C network analyzer was used after first being calibrated. The measured results have average insertion loss $|S_{21}|$ of 0.94 dB, average return loss $|S_{11}|$ of 16.2 dB and 3-dB fractional bandwidth (FBW) = 52.7%. The transmission zeros appeared near the passband at 2.5 GHz (TZ_1), 2.95 GHz (TZ_2) and 6.1 GHz (TZ_3), thus having a high isolation. The measured group delay has a small varying range from 0.2 ns to 0.4 ns. Although the simulated results and the measured results have a slight mismatch in the high frequency region, this is typically caused due to the material limitation of the substrate. The filter has a simple structure and good filter features, thus it has a good potential to be applied in UWB applications.

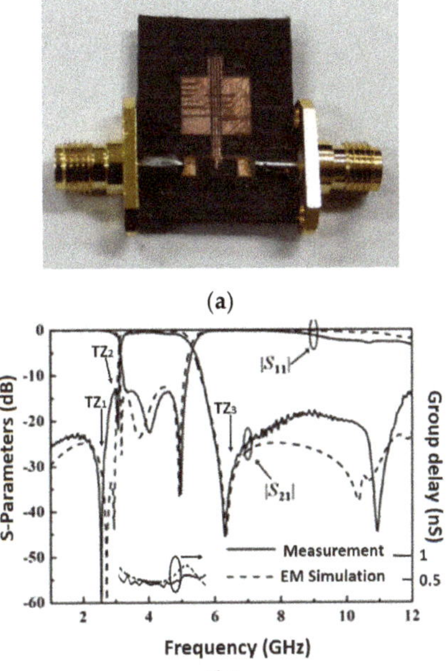

Figure 5. (a) Photograph and (b) simulated and measured results of the BPF with a wide stopband. $L_1 = 3.35$ mm, $L_2 = 7.1$ mm, $L_s = 13.7$ mm, $W_1 = 3.38$ mm, $W_2 = 0.2$ mm, $W_s = 0.2$ mm, $G_1 = 0.2$ mm, $G_2 = 0.15$ mm, $G_3 = 0.15$ mm.

4. Design of Dual UWB Bandpass Filter

In the second design example, a dual ultra-wideband BPF is presented. Figure 6 shows the configuration of the dual UWB BPF. This filter combines two UWB BPFs based on the above work of section III. The upper SLSIR is used to form the first UWB BPF with 3.1 to 5 GHz and the lower SLSIR is used to form the second UWB BPF with 6 to 10.2 GHz. Before combining the first UWB BPF and the second UWB BPF, they are designed independently to match the requirements of the first and second UWB responses. As addressed above, in the used SLSIR, the first odd mode and first even mode constitute together a passband of the ultra-wideband. The passband edge of filter has five transmission zeros, also providing a good attenuation and isolation. Moreover, the designed dual UWB BPF can be seen as a UWB BPF with a notch band.

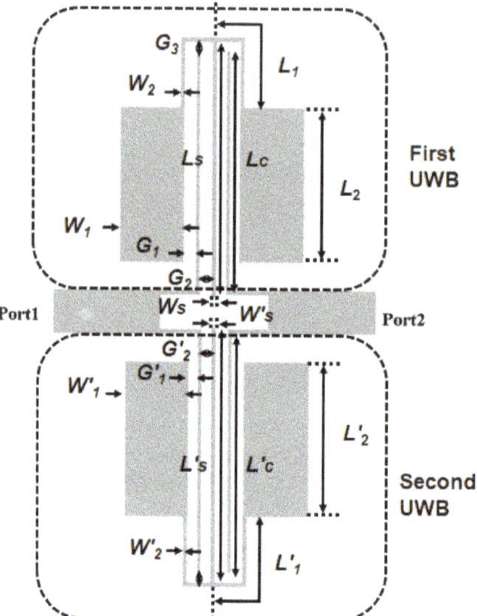

Figure 6. Configuration of the dual UWB BPF.

4.1. Design Procedure

The first UWB response is set from 3.1 to 5 GHz with a stopband from 5.5 to >11 GHz, as in Section 3. The designed UWB BPF with a wide stopband is optimized using the EM simulation [30] and the dimension parameters are shown in Figure 4. The simulated results have a wide stopband with an attenuation is larger than 20 dB from 5.9 to 12 GHz, as shown in Figure 4. Therefore, the second UWB response can be designed in the wide stopband region of the first UWB response.

The second UWB response is thus set from 6 to 10.1 GHz. Using the same structure as shown in Figure 4, the desired resonant frequencies of the SLSIR are set as: f_{o1} = 6 GHz and f_{e1} = 9.3 GHz. Namely, the passband frequency ratio f_{e1}/f_{o1} is 1.55. Similarly, there are many ways to match the requirements. To simplify the design, the groups (K_1, K_2) can be determined at the same values for the first UWB response and the second UWB response. In this study, K_1 = 0.25 and K_2 = 1, as mapped in Figure 2a, are used and then the corresponding length ratio R_1 is found to be 0.7 and length ratio R_2 is found to slightly more than 0.9, where the passband frequency ratios of f_{e1}/f_{o1}= 1.58 and f_{o2}/f_{o1} greater than 4.64. The transmission zeros are calculated to be 5.2 GHz (TZ$_3$) by using (3a), and the transmission zeros are determined to be 6.1 GHz (TZ$_5$) and 12.2 GHz (TZ$_6$) by using (3b).

The second UWB BPF is designed and optimized using a full wave EM simulation [30] and the dimension parameters are: L_1' = 1.85 mm, L_2' = 3.1 mm, L_s' = 6.95 mm, L_c' = 5.92 mm, W_1' = 3.43 mm, W_2' = 0.2 mm, W_s' = 0.2 mm, G_1' = 0.2 mm, G_2' = 0.15 mm, G_3' = 0.15 mm. Figure 7 displays the simulated results which have insertion loss |S$_{21}$| of 0.67 dB return loss |S$_{11}$| of 13.81 dB and 3-dB fractional bandwidth (FBW) = 53%. Three transmission zeros are also appeared near the passband to have a high band selectivity.

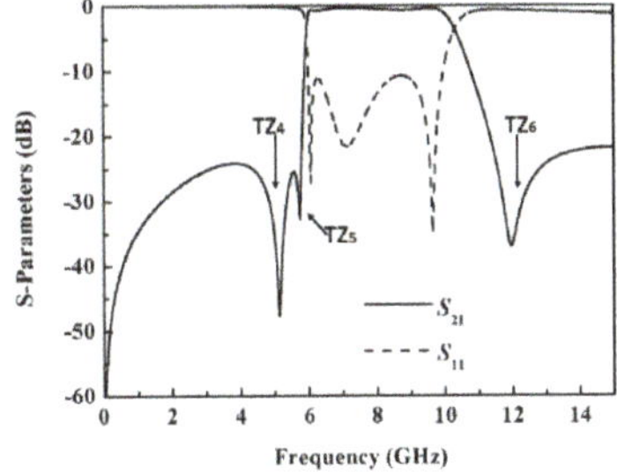

Figure 7. Simulated results of the second UWB BPF using EM simulation. For the second UWB BPF, $L_1' = 1.85$ mm, $L_2' = 3.1$ mm, $L_s' = 6.95$ mm, $L_c' = 5.92$ mm, $W_1' = 3.43$ mm, $W_2' = 0.2$ mm, $W_s' = 0.2$ mm, $G_1' = 0.2$ mm, $G_2' = 0.15$ mm, $G_3' = 0.15$ mm.

4.2. Measured Results

After designing the first UWB response and the second UWB response independently, they were combined together as shown in Figure 6, and the simulated results shown in Figure 8b include the individual response of the lower UWB shown in Figure 4 and higher UWB shown in Figure 7. The filter is also realized on the same commercial substrate (RT/Duroid 5880). Figure 8a displays the photograph of the fabricated BPF. The filter has length of 0.15 λ_g and width of 0.40 λ_g, where λ_g is the guided wavelength at the first center frequency. The filter is fabricated and measured by an HP8510C Network Analyzer. Figure 8b displays simulated and measured results of the UWB BPF of the first UWB response from 3.1 to 5 GHz and second UWB response from 6 to 10.1 GHz. The measured results have average insertion loss $|S_{21}|$ of 1.3 dB, average return loss $|S_{11}|$ of 10 dB and 3-dB fractional bandwidth (FBW) = 52.1%, for the first UWB response; and average insertion loss $|S_{21}|$ of 1.15 dB, average return loss $|S_{11}|$ of 15 dB and 3-dB fractional bandwidth (FBW) = 55.7%, for the second UWB response. The transmission zeros appeared close the passband at 2.5 GHz (TZ_1), 2.95 GHz (TZ_2), 5.3 GHz (TZ_4), 6.2 GHz (TZ_3, TZ_5) and 11.3 GHz (TZ_6) to have a high band selectivity. The transmission zeros (TZ_3, TZ_5) are located at the same frequency and then overlapped. Measured group delays show an acceptable value varying from 0.2 ns to 1 ns. The proposed dual UWB BPF also has simple structure and good filter features, showing a good potential to be applied in UWB applications.

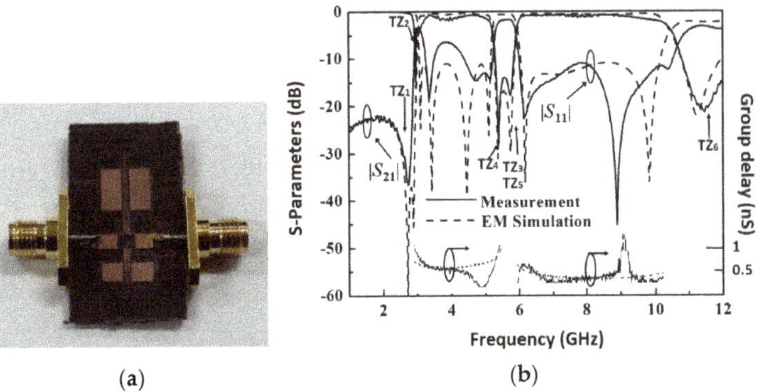

(a) (b)

Figure 8. (a) Photograph and (b) simulated and measured results of the UWB BPF and group delay of the first UWB response from 3.1 to 5 GHz and second UWB response from 6 to 10.1 GHz. For the first UWB BPF, $L_1 = 3.35$ mm, $L_2 = 7.1$ mm, $L_s = 13.7$ mm, $W_1 = 3.38$ mm, $W_2 = 0.2$ mm, $W_s = 0.2$ mm, $G_1 = 0.2$ mm, $G_2 = 0.15$ mm, $G_3 = 0.15$ mm. For the second UWB BPF, $L_1' = 1.85$ mm, $L_2' = 3.1$ mm, $L_s' = 6.95$ mm, $W_1' = 3.43$ mm, $W_2' = 0.2$ mm, $W_s' = 0.2$ mm, $G_1' = 0.2$ mm, $G_2' = 0.15$ mm, $G_3' = 0.15$ mm.

Table 1 shows comparison of filter performances of the designed prototype with the previous published works, using parallel-coupled lines [10], asymmetric SIR [12], embedded open stubs [15], three pairs of coupled line sections with open-circuited stubs [23] and multi-layered substrate with two stub-loaded resonators [24]. This work has more transmission zeros and thus higher band selectivity than the conventional designs. Further, the fabricated dual-UWB BPF can be very compact. The measured results closely match the simulated results.

Table 1. Comparison of filter performances from this work with the previous published designs.

	Ref. [8]	Ref. [9]	Ref. [12]	Ref. [15]	Ref. [23]	Ref. [24]	This Work 1	This Work 2
Center frequency (GHz)	3.97	4/ 8	4	6.5	1	2.3	4.2	4.05/ 8.05
$|S_{11}|$ (dB)	15	10/12	15	17	15	13	16.5	8/12
$|S_{21}|$ (dB)	0.8	1.3/2.4	0.5	0.5	1.0	0.35	0.94	1.3/1.1
3-dB FBW (%)	45.3	42/58	62.5	113	123	80	52.7	52/55
Circuit size ($\lambda g \times \lambda g$)	0.66×0.43	0.70×0.28	1.12×0.1	0.4×0.06	0.17×0.14	0.53×0.43	0.15×0.25	0.15×0.4
Number of notched bands	0	1	0	1	0	2	0	1
Numbers of transmission zeros	2	2	3	0	1	1	3	5

5. Conclusions

This paper reported a stub-loaded stepped impedance resonator (SLSIR) and analyzed the even and odd resonant modes of this resonator using the input impedance/admittance analysis. For controlling the resonant modes of the SLSIR, two impedance ratios and two length ratios can be used as the design parameters. Several resonant mode variation curves were developed. Using resonant mode variation curves, resonant modes of the SLSIR can be tuned by the desired impedance ratios and length ratios to form different filtering types. In this paper, two types of the UWB BPFs were presented: a UWB filter with a wide stopband and a dual UWB filter (namely, a UWB with a notch band). The first odd mode and first even mode are selected to constitute an ultra-wideband, and the other odd mode and even mode are moved to higher frequencies. Moreover, since the input and output ports are set into the folded SLSIR, several transmission zeros are generated for the type of the resonator. The transmission zeros are generated for the coupling length between the coupled line with the resonator, and generated by the open stub. The transmission zeros are generated, improving the band selectivity of the designed filter using the SLSIR. The slight error of the band response is generated due to the substrate error. The simulation and measurement of the two designed cases are almost consistent, thus verifying the proposed design concepts.

Author Contributions: M.-H.W. contributed the design concept and wrote the paper, F.-Z.Z. and H.-Z.L. simulated, fabricated and measured the filter, and S.-K.L. provided the design resource and advised on the design procedure. All authors have read and agreed to the published version of the manuscript.

Funding: This research received no external funding.

References

1. US Federal Communications Commission. Revision of Part 15 of the Commission's Rules Regarding Ultra-Wideband Transmission Systems. In *First Report and Order, ET Docket*; Federal Communications Commission: Washington, DC, USA, 2002.
2. Weng, M.H.; Liu, S.K.; Liu, Y.T.; Hung, C.Y. A low loss ring UWB bandpass filter. *Microw. Opt. Tech. Lett.* **2008**, *50*, 2317–2319. [CrossRef]

3.   Kim, C.H.; Chang, K. Ultra-wideband (UWB) ring resonator bandpass flter with a noched band. *IEEE Microw. Wirel. Compon. Lett.* **2011**, *21*, 206–208. [CrossRef]

4.   Lan, S.W.; Weng, M.H.; Hung, C.Y.; Chang, S.J. Design of a compact ultra-wideband bandpass filter with an extremely broad stopband region. *IEEE Microw. Wireless Compon. Lett.* **2016**, *26*, 392–394. [CrossRef]

5.   Wang, H.; Kang, W.; Miao, C.; Wu, W. Cross-shaped UWB bandpass filter with sharp skirt and notched band. *Electron. Lett.* **2012**, *48*, 96–97. [CrossRef]

6.   Lee, J.K.; Kim, Y.S. Ultra-wideband bandpass filter with improved upper stopband performance using defected ground structure. *IEEE Microw. Wirel. Compon. Lett.* **2010**, *20*, 316–318. [CrossRef]

7.   Song, Y.; Yang, G.M.; Geyi, W. Compact UWB bandpass filter with dual notched bands using defected ground structures. *IEEE Microw. Wirel. Compon. Lett.* **2014**, *24*, 230–232. [CrossRef]

8.   Weng, M.H.; Huang, C.Y.; Hung, C.Y.; Lan, S.W. Design of an ultra-wideband bandpass filter by using coupled three line microstrip structure. *J. Electromagn. Waves Appl.* **2012**, *26*, 716–728. [CrossRef]

9.   Weng, M.H.; Hsu, C.W.; Lan, S.W.; Yang, R.Y. An ultra-wideband Bbndpass filter with a notch band and a wide upper bandstop performances. *Electronics* **2019**, *8*, 1316. [CrossRef]

10.   Bojra, A.L.; Belenguer, A.; Cascon, J.; Esteban, H.; Boria, V.E. Wideband passband transmission line based on metamaterial-inspired CPW balanced cells. *IEEE Antenna Wirel. Propag. Lett.* **2011**, *10*, 1421–1424.

11.   Hameed, M.; Xiao, G.; Najam, A.I.; Qiu, L.; Hameed, T. Quadruple-mode wideband bandpass filter with improved out-of-band rejection. *Electronics* **2019**, *8*, 300. [CrossRef]

12.   Chang, Y.C.; Kao, C.H.; Weng, M.H.; Yang, R.Y. Design of compact wideband bandpass filter using asymmetric SIR with low loss, high selectivity and wide stopband. *IEEE Microw. Wirel. Compon. Lett.* **2008**, *18*, 770–772. [CrossRef]

13.   Ghatak, R.; Sarkar, P.; Mishra, R.K.; Poddar, D.R. A compact UWB bandpass filter with embedded SIR as band notch structure. *IEEE Microw. Wirel. Compon. Lett.* **2011**, *21*, 261–263. [CrossRef]

14.   Li, R.; Zhu, L. Compact UWB bandpass filter using stub-loaded multiple-mode resonator. *IEEE Microw. Wirel. Compon. Lett.* **2007**, *17*, 40–42. [CrossRef]

15.   Weng, M.H.; Liauh, C.T.; Wu, H.W.; Vargas, S.R. An ultra-wideband bandpass filter with an embedded open-circuited stub structure to improve in-band performance. *IEEE Microw. Wirel. Compon. Lett.* **2009**, *19*, 146–148. [CrossRef]

16.   Zhou, Y.; Yao, B.; Cao, C.; Deng, H.; He, X. Compact UWB bandpass filter using ring open stub loaded multiple-mode resonator. *Electron. Lett.* **2009**, *45*, 554–556. [CrossRef]

17.   Chu, Q.X.; Tian, X.K. Design of UWB bandpass filter using stepped-impedance stub-loaded resonator. *IEEE Microw. Wirel. Compon. Lett.* **2010**, *20*, 501–503. [CrossRef]

18.   Han, L.; Wu, K.; Zhang, X. Development of packaged ultrawideband bandpass filters. *IEEE Trans. Microw. Theory Tech.* **2010**, *58*, 20–228.

19.   Nosrati, M.; Mirzaee, M. Compact wideband microstrip bandpass filter using quasi-spiral loaded multiple-mode resonator. *IEEE Microw. Wirel. Compon. Lett.* **2010**, *20*, 607–609. [CrossRef]

20.   Tu, W.H. Sharp-rejection broadband microstrip bandpass filter using penta-mode resonator. *Electron. Lett.* **2010**, *46*, 772–773. [CrossRef]

21.   Fan, J.; Zhan, D.; Jin, C.; Luo, J. Wideband microstrip bandpass filter based on quadruple mode ring resonator. *IEEE Microw. Wirel. Compon. Lett.* **2012**, *22*, 348–350. [CrossRef]

22.   Kumar, S.; Gupta, R.D.; Parihar, M.S. Multiple band notched filter using C-shaped and E-shaped resonator for UWB applications. *IEEE Microw. Wirel. Compon. Lett.* **2016**, *26*, 340–342. [CrossRef]

23.   Zhang, T.; Xiao, F.; Bao, J.; Tang, X.H. Compact ultra-wideband bandpass filter with good selectivity. *Electron. Lett.* **2016**, *52*, 210–212. [CrossRef]

24.   Li, Y.; Choi, W.W.; Tam, K.W.; Zhu, L. Novel wideband bandpass filter with dual notched bands using stub-loaded resonators. *IEEE Microw. Wirel. Compon. Lett.* **2017**, *27*, 25–27.

25.   Velázquez-Ahumada, M.D.C.; Martel-Villagr, J.; Medina, F.; Mesa, F. Application of stub loaded folded stepped impedance resonators to dual band filters. *Prog. Electromagn. Res.* **2010**, *102*, 107–124. [CrossRef]

26.   Liu, S.K.; Zheng, F.-Z. A new compact tri-band bandpass filter using step impedance resonators with open stubs. *J. Electromagn. Waves Appl.* **2012**, *26*, 130–139. [CrossRef]

27.   Makimoto, M.; Yamashita, S. Bandpass filter using parallel-coupling stripline stepped impedance resonators. *IEEE Trans. Microw. Theory Tech.* **1980**, *28*, 1413–1417. [CrossRef]

28.   Pozar, D.M. *Microwave Engineering*, 2nd ed.; John Wiley & Sons, Inc.: Hoboken, NJ, USA, 1998.

29. Hong, J.S.; Lancaster, M.J. *Microstrip Filters for RF/Microwave Applications*; John Wiley & Sons, Inc.: Hoboken, NJ, USA, 2001.
30. Zeland Software, Inc. *IE3D Simulator*; Zeland Software, Inc.: Fremont, CA, USA, 2002.

Article

A Miniaturized Wideband Bandpass Filter Using Quarter-Wavelength Stepped-Impedance Resonators

Liqin Liu [1], Ping Zhang [1], Min-Hang Weng [1], Chin-Yi Tsai [2] and Ru-Yuan Yang [2,*]

[1] School of Information Engineering, Putian University, Putian 351100, Fujian, China; llqsmile006@gmail.com (L.L.); zhangpingdudu@gmail.com (P.Z.); hcwweng@gmail.com (M.-H.W.)

[2] Graduate Institute of Materials Engineering, National Pingtung University of Science and Technology, Pingtung County 912, Taiwan; tylertsaiji@gmail.com

* Correspondence: ryyang@mail.npust.edu.tw; Tel.: +886-8-7703202

Received: 22 October 2019; Accepted: 10 December 2019; Published: 13 December 2019

Abstract: In this paper, we present a simple method to design a miniaturized wideband bandpass filter with suppression of the third harmonic, using only two quarter-wavelength stepped-impedance resonators (SIRs). The resonant modes of the quarter-wavelength SIR, depending on the impedance ratio (K) and electrical length ratio (α), are discussed first. As to setting the resonant frequency of the SIR for the lower band edge of the required band, the size parameters of two quarter-wavelength SIRs can be determined by selecting the desired impedance ratio (K) and length ratio (α). By using the opposite directional arrangement of two SIRs with direct taped input/output ports, the wideband response can be formed. A filter example is shown in this study to address this simple design procedure. The measured results of the fabricated filter have a wide passband response from 3.3 to 5.8 GHz, with an insertion loss of 1.5 dB, a return loss of 20 dB, an extended bandwidth ration of 55%, a low-average group delay of less than 0.75 ns, and a stopband from 6 to 12 GHz, with an attenuation level of 20 dB. Due to the similar 0° feeding, a transmission zero at 8.3 GHz appears near the band edge; thus, improving the band selectivity. The proposed filter can have a very simple structure and a miniature size. Simulated results and measured results are in good agreement.

Keywords: wideband; bandpass filter; quarter wavelength; stepped-impedance resonator (SIR)

---

1. Introduction

In the last 20 years, communications systems have developed rapidly. The bandpass filter (BPF) used in the radio-frequency (RF) front end is an important device for selecting the desired signals for the use of the communications system [1]. Typically, the requirements of the filter comprise low passband loss, sharp band selectivity, spurious suppression, a compact size, and a low cost. The wideband system has been rapidly expanding ever since in 2002 the U.S. Federal Communications Commission (FCC) approved the unlicensed applications of ultra-wideband (UWB) with frequency ranging from 3.1 to 10.6 GHz for several uses, such as hand-held and indoor systems [2]. The direct sequence ultra-wideband (DS-UWB) specifications for wireless personal area networks (WPANs) is further divided into a low band of 3.1–5.1 GHz and a high band of 6.2–9.7 GHz, to avoid the frequent use of IEEE 802.11a wireless local area networks (WLANs) at 5–6 GH. As one of the important component blocks, some wideband BPFs were developed to obtain the desired fractional bandwidth (FBW) [3–15].

In Chang's work [3], a wide BPF adopted a combination of serial and shunt microstrip units to have an FBW of 54%. However, the drawbacks of this filter are the large size and complex structure. In Hung's work [4], a wideband filter using parallel coupled lines was reported to obtain an FBW of 80%, with a complex image impedance design method. Various types of multiple-mode resonators (MMRs) were used to obtain the wideband filter. In Zhu's work [5,6], the resonant modes of the MMR were analyzed first, and then controlled to be coupled together to obtain the desired bandwidth. In

Killamsetty's work [7], a short-circuited loaded triangular stub resonator was used to obtain a wideband BPF, but the structure was still complex. In Wang's work [8] and Ye's work [9], ultra-wideband (UWB) BPFs were designed by semi-lumped or hybrid microstrip/coplanar waveguide (CPW) structures. In Chang's work [10], a stepped-impedance resonator (SIR) was developed to realize a wideband response, which showed a high band selectivity. In Choudhary's work [11], split circular rings and a rectangular stub were used to design a via-less metamaterial wideband BPF, but the bandwidth was insufficiently large. In Ji's work [12], a multilayer structure was used to achieve a wideband BPF. In Li's work [13], series and shunt resonators were coupled together to directly realize a wideband BPF. In Li's work [14], a single wavelength ring SIR was used to implement a wide-frequency band. In Gao's work [15], a combination of open/shorted stubs was used to obtain a wideband response with an improved upper-stopband. In Hameed's work [16], multiple-mode split-ring resonators in a waveguide cavity were designed to obtain a wideband BPF. However, the above design procedure and the device structure are complex.

In this study, a simple method is reported to design a wideband BPF with an FBW greater than 50%, and with suppression of the third harmonic. Only two quarter-wavelength SIRs are needed in this design. The design concept and procedure are described in this study. The designed filter can simultaneously have a simple structure and miniature size. To prove the design concept, a filter example is presented. The measured results of the fabricated filter match with the simulated results, showing that the filter can be suitable for a practical RF system.

2. Design Procedure

Figure 1 displays the structure of the proposed wide BPF. The basic element of the filter is two microstrip quarter-wavelength SIRs. (L_1, L_2) and (W_1, W_2) are the physical lengths and widths of the high impedance section and low impedance section of SIR 1, respectively. (L_3, L_4) and (W_3, W_4) are the physical lengths and widths of the high impedance section and low impedance section of SIR 2, respectively. Gap (g) is the spacing between SIR 1 and SIR 2. Expression (p) is the physical length from the input/output ports to the short ends of SIR 1 and SIR 2. For this design, a low-cost FR4 substrate is used, having a dielectric constant (ε_r) of 4.4, a loss tangent $(\tan\delta)$ of 0.02, and a thickness of 1.6 mm.

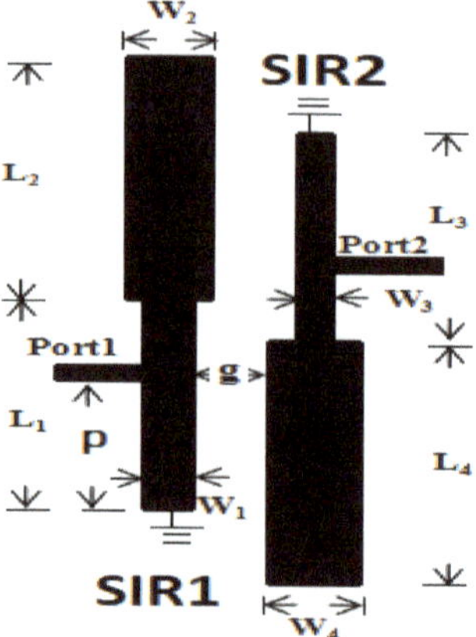

Figure 1. The layout of the proposed wideband bandpass filter (BPF).

Analysis of Quarter-Wavelength Stepped-Impedance Resonator

Figure 2a displays the construction of the quarter-wavelength SIR. The SIR is formed by an impedance unit (Z_1) with an electrical length (θ_1), and an impedance unit (Z_2) with an electrical length (θ_2). The impedance ratio of this quarter-wavelength SIR is defined as $K = Z_2/Z_1$ and the electrical length ratio is defined as $\alpha = \theta_2/(\theta_1+\theta_2) = \theta_2/\theta_t$. The characteristics of the conventional SIR are able to control the higher-order resonance modes, closer or far away, efficiently, with a different impedance ratio (K) and electrical length ratio (α). The input impedance (Z_{in}) of the quarter-wavelength SIR is derived as [17]:

$$Z_i = jZ_2 \frac{Z_1 \tan\theta_1 + Z_2 \tan\theta_2}{Z_2 - Z_1 \tan\theta_1 \tan\theta_2}. \tag{1}$$

The resonant condition can be obtained when $Y_{in} = 0$ as follows [1]:

$$\tan\theta_1 \tan\theta_2 = K \tag{2}$$

To reduce the design parameters, θ_1 and θ_2 are derived by α and θ_t as follows:

$$\theta_1 = (1 - \alpha) \cdot \theta_t \tag{3}$$

$$\theta_2 = \alpha \cdot \theta_t \tag{4}$$

Thus, the resonant condition controlled by the impedance ratio (K) and the electric length ratio (α) is expressed as:

$$K = \tan[(1 - \alpha) \cdot \theta_t] \cdot \tan(\alpha \cdot \theta_t). \tag{5}$$

Figure 2b illustrates the resonant condition curve of the quarter-wavelength SIR. As shown in Figure 2b, the total electrical length becomes shorter and longer when the impedance ratio (K) is smaller and larger than 1, respectively. In other words, for the same resonator size, the frequency of the quarter-wavelength resonator shifts lower and higher when the impedance ratio (K) is smaller and larger than 1, respectively [17].

Figure 3 illustrates (a) the structure of the quarter-wavelength SIR with a special case of K = 1 and (b) the simulated frequency responses with different feeding positions of the input and output ports. For the special case of K = 1, the quarter-wavelength SIR is seen as the conventional quarter-wavelength uniform impedance resonator (UIR). Based on the coupled line theory, the conventional quarter-wavelength UIR with opposite directions would have a passband response. The bandwidth of the passband can be controlled by the impedance and gap of the coupled lines, namely the even mode impedance Z_{even} and odd mode impedance Z_{odd} of the coupled lines [18]. Moreover, there are many harmonics that appeared in the higher frequency. The frequency of the first harmonic is 3 times the fundamental mode, as shown in Figure 2c. Moreover, it is known that no matter what value α is, the fundamental resonant mode is kept and resonant, as $(\theta_1 + \theta_2) = \theta_t$ is equal to the quarter-wavelength, as shown in Figure 2b.

It is known that the feeding positions (t) of the input and output (I/O) ports on the resonators affect the filter performances. Figure 3b shows the simulated filter responses of the quarter-wavelength UIR with different feeding positions (t) of the input and output (I/O) ports, where t is the physical length from the center of the quarter-wavelength UIR to the short end. The simulation was done by using the full-wave electro-magnetic simulator IE3D [19]. In this simulation, $L_1 + L_2$ and $L_3 + L_4$ were both selected and kept as 11 mm to be the quarter-wavelength at 4 GHz. The impedance of the UIR was 100 Ω and the gap (g) was 2 mm. It was clearly found that as the feeding positions of the I/O ports are separated away, there were two modes that existed in the resonator. As shown in Figure 3b,c, when t is increased from 0 mm to 1 mm, a wideband response is formed and two separated modes appear. However, as t is further increased from 1 mm to 4 mm, the band response is degraded since the modes are separated and not coupled together. The first resonant mode at lower frequency is the fundamental mode of the quarter-wavelength UIR. Since the open stub can also be seen as the quarter-wavelength resonator, the other mode at higher frequency is resonant at the length from the I/O ports to the open end ($t + L_2$ (L_4)). Namely, the frequency of the other mode can be roughly estimated by seeing that the length ($t + L_2$ (L_4)) is a quarter-wavelength. Therefore, by carefully selecting the feeding positions of the I/O ports, the desired passband response can be formed. At t = 1 mm, the bandwidth of the passband is from 4 to 6.2 GHz, namely, the FBW is 43%. However, it is also known that the next harmonic also appeared at around 12 GHz, which is 3 times the fundamental frequency of the quarter-wavelength resonator.

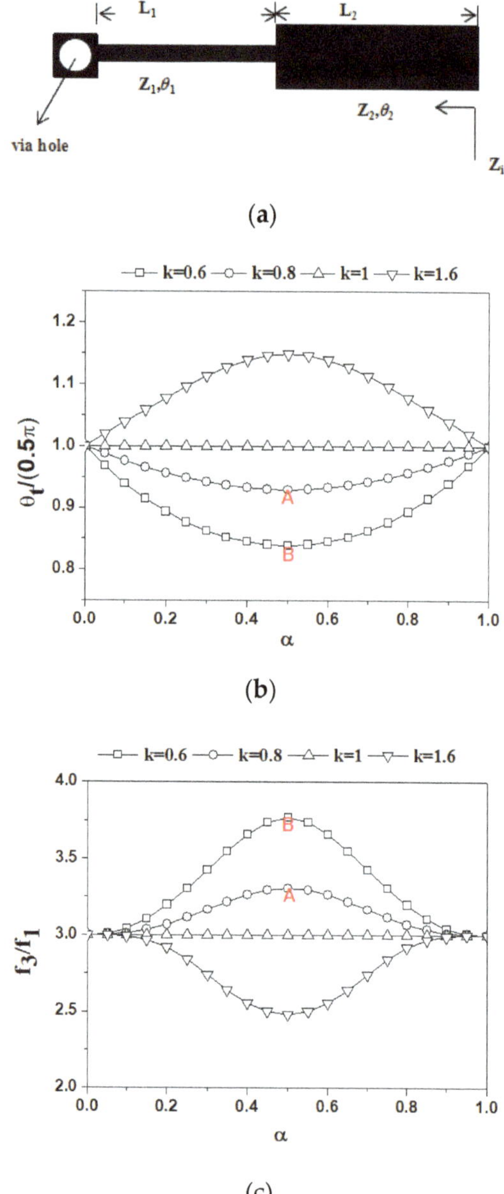

Figure 2. (**a**) The construction and (**b**) resonant condition curve (**c**) f_3/f_1 of the quarter-wavelength stepped-impedance resonator (SIR).

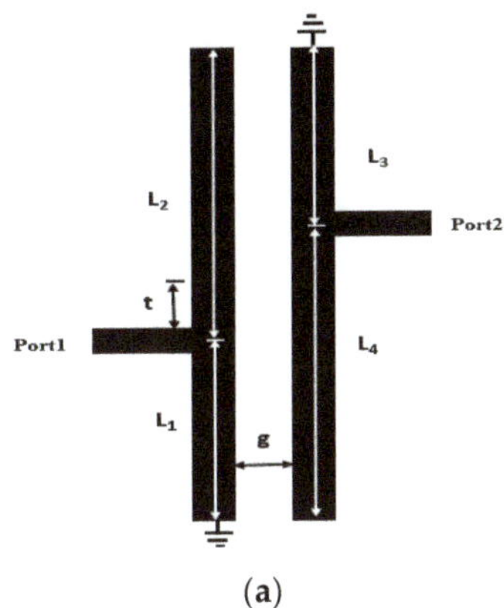

(**a**)

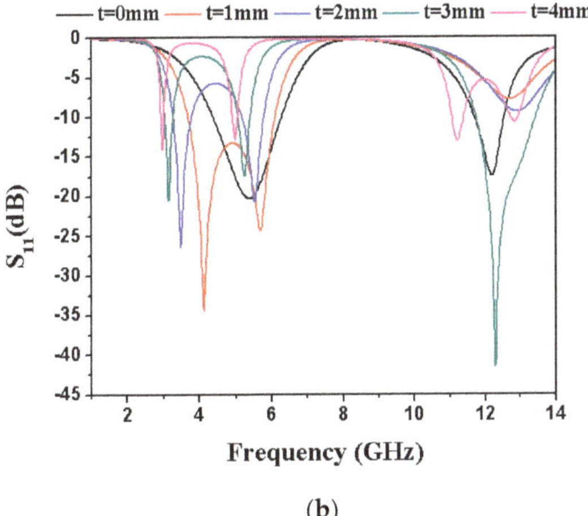

(**b**)

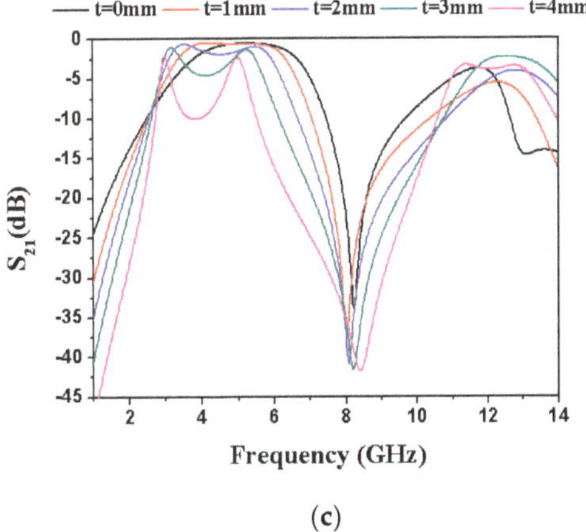

(**c**)

Figure 3. (**a**) the structure of the quarter-wavelength SIR with special case of K = 1 and (**b**) the simulated frequency responses with different feeding positions (**c**) of the input and output ports.

Figure 4 shows the current distribution of the quarter-wavelength UIR at 3.2, 3.6, 5.2, and 5.6 GHz. In this simulation, t was 1 mm, and a wideband response with an FBW of 43% was obtained, as shown in Figure 3c. Typically, the current distribution of the filter at the resonant frequency is used to show the locations of the maximum and minimum electromagnetic energy, namely, to know where resonance occurs in the structure. This resonant mode can be then excited by providing the suitable input and output terminals in the resonant excitation location. The resonant mode can be suppressed if a dispersing device is used in the resonant excitation location to avoid the resonant mode. As shown in Figure 4, the resonant energy at lower frequency (see 3.2 GHz and 3.6 GHz) is distributed over the UIR, and the resonant energy at higher frequency (see 5.2 GHz and 5.6 GHz) is distributed mostly on the area from the I/O ports to the open end (t+ L_2 (L_4)).

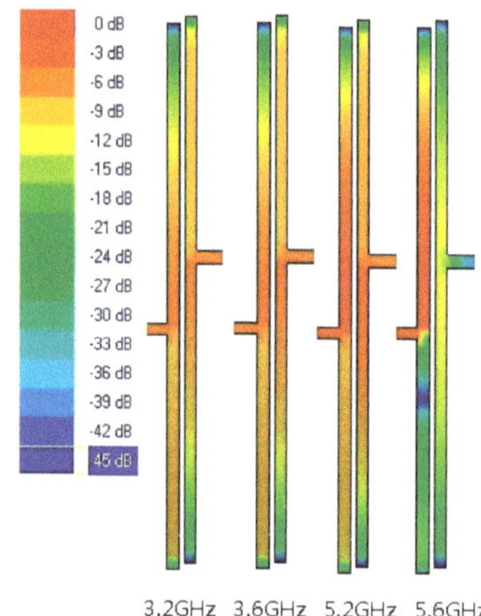

Figure 4. Current distribution of the quarter-wavelength uniform impedance resonator (UIR) at 3.2, 3.6, 5.2, and 5.6 GHz.

Based on the above discussion, in this study, to further extend the bandwidth of the wideband filter, the quarter-wavelength UIR (K = 1) was replaced with the quarter-wavelength SIR. The filter example is designed around 4.2 GHz. The resonant frequencies of the quarter-wavelength SIR were selected, first based on Figure 2b,c, and then on a wide bandpass response that can easily be achieved by coupling the resonant modes with a careful arrangement of the I/O ports. As mapping spot A and spot B in Figure 2b for SIR 1 and SIR 2, respectively, (K = 0.8, α = 0.5) and (K = 0.6, α = 0.5) were selected in this design for miniaturizing the filter size and providing two lower resonant modes at 3.9 GHz and 3.5 GHz, respectively. Therefore, for the quarter-wavelength SIR 1, the physical width and length are 0.35 mm (W_1) and 5 mm (L_1), and 1.1 mm (W_2) and 5 mm (L_2) for the high impedance section (Z_1 = 100 Ω), and the low impedance section (Z_2 = 60 Ω), respectively. For the quarter-wavelength SIR 2, the physical width and length are 0.35 mm (W_3) and 5 mm (L_3), and 0.6 mm (W_4) and 5 mm (L_4) for the high impedance section (Z_1 = 100 Ω) and the low impedance section (Z_2 = 80 Ω), respectively. The reason for using two different SIRs is to extend the bandwidth of the designed filter. Moreover, as shown in Figure 2c, the frequency of the third harmonic mode (f_3) over the frequency of fundamental mode (f_1), namely f_3/f_1, is moved to higher values of 3.3 and 3.7 for SIR 1 and SIR 2, as mapping the spot A and spot B in Figure 2c, respectively. Thus, the third harmonics of this filter are suppressed, since the two third harmonics of the two different SIRs (SIR1 and SIR2) are different, and cannot be coupled.

After determining the two quarter-wavelength SIRs, the input/output (I/O) ports were directly taped onto the quarter-wavelength SIRs. To achieve good external quality, and a transmission zero, a similar 0° feed structure of the I/O ports was directly taped to the two quarter-wavelength SIRs at p = 3 mm [1]. As discussed above, in this wide passband, other higher resonant modes are also excited. When p = 3 mm, the two quarter-wavelength SIRs have total physical lengths of 7 mm from the feeding position of the I/O ports to the open ends, and would provide two other modes at 5.1 GHz and 5.6 GHz, respectively, as also mapping into Figure 2b.

Figure 5 shows the simulated filter responses when varying the coupling gap (g) between two quarter-wavelength SIRs. It was found that as the g value decreases from 0.5 to 0.05 mm, the passband became wider and the insertion loss became lower. Because the minimum carving size of the carving machine is 0.15 mm, the coupling gap, g = 0.15 mm, was used to obtain the maximum passband with a low insertion loss of 1.2 dB, a return loss of 20 dB, and a 3 dB FBW of 55% (from 3.3 to 5.8 GHz). Moreover, by taking advantage of the direct 0° feed structure of the I/O ports, a transmission zero at 8.3 GHz was generated near the passband edge in the filter response [20]. In addition, it was clearly observed that the third harmonic of the quarter-wavelength UIR near 12 GHz was suppressed; thus, a stopband with an attenuation of 15 dB from 7.0 to 12 GHz was obtained.

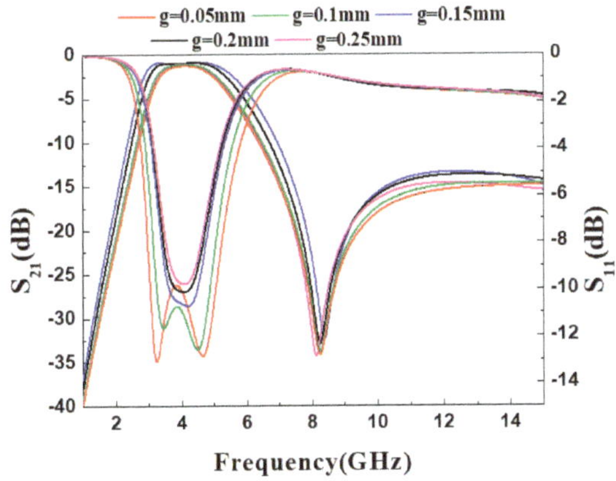

Figure 5. Filter responses when varying the coupling gap (g) between two quarter-wavelength SIRs.

Figure 6 further shows the current distribution of the proposed wideband filter at the frequencies of 3.5, 4.0, 5.5, and 6 GHz. As shown in Figure 6, the resonant energy at lower frequency (see 3.5 GHz and 4.0 GHz) is distributed over the SIR, and the resonant energy at higher frequency (see 5.5 GHz and 6.0 GHz) is distributed mostly on the areas from the I/O ports to the open end, thus verifying the design concept.

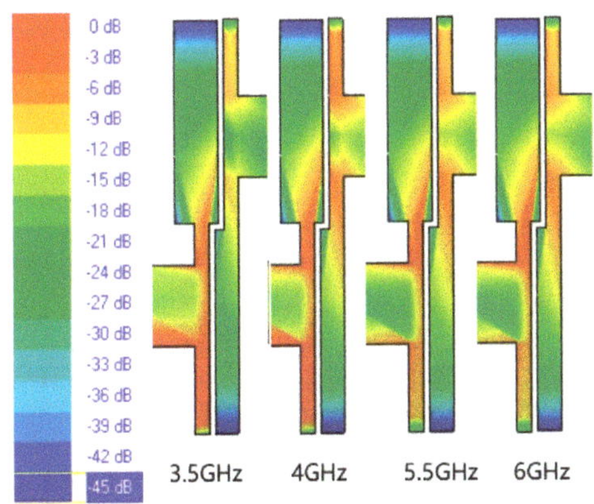

Figure 6. Current distribution of the proposed wideband filter at lower and higher frequencies.

With this simple method and design concept, the resonant frequencies of the quarter- wavelength SIR are selected first, and then a wide bandpass response can be easily achieved by coupling the resonant modes with a careful arrangement of the I/O ports.

3. Experimental Results

The filter sample was fabricated using conventional printing circuit board technology. Figure 7a shows a photograph of the fabricated sample. The whole size of the fabricated filter is 12 mm X 4 mm, i.e., approximately 0.3 λg by 0.1 λg, where λg is the guided wavelength at the center frequency. Measurement was processed by an HP8722ES network analyzer. Before measurement, two coaxial cables of the network analyzer, which were connected to the I/O ports of the fabricated filter sample, were carefully calibrated by using short-open-load-through calibration. Steps were carefully processed to make sure that the S_{21} was close to zero when the two coaxial cables were connected to the load-through device. The measured results shown in Figure 7b exhibit a center frequency of 4.2 GHz with a low insertion loss of 1.2 dB over the passband, a return loss greater than 15 dB, a 3 dB FBW of 55% (from 3.3 to 5.8 GHz), and a stopband with an attenuation of 15 dB, from 7.5 to 12 GHz. Moreover, the transmission zero at 8.3 GHz was clearly obtained because of the use of a 0° feeding structure [20]; thus, a good band selectivity was also achieved. The group delay was obtained by taking the derivative of the phase. Figure 7c shows that the average calculated group delay of the fabricated filter is less than 0.75 ns over the whole passband. As compared to other works with a group delay, this group delay of this design is acceptable. The simulated results and the measured results are mostly in agreement, with a slight mismatch in the high band edge of the passband. This mismatch may have been due to the fact that the electromagnetic phenomenon of the solder at the short-circuited end was not well-considered, or the amount of solder at the short-circuited end was not appropriate.

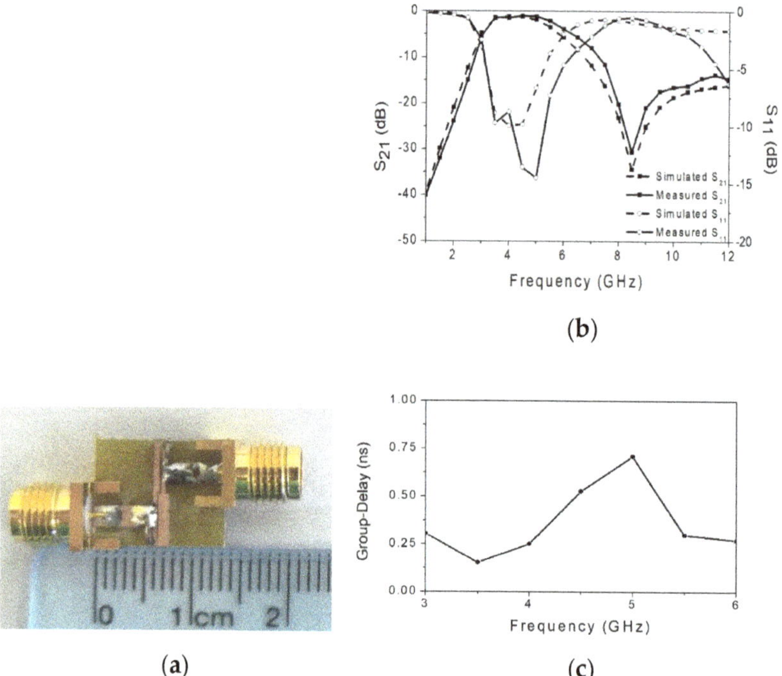

Figure 7. (a) A photograph of the fabricated sample, (b) measured filter responses, and (c) the calculated group delay of the fabricated filter.

Table 1 compares this design to some reported works. The designed filter shows acceptable filter performance when compared to other filters. In addition, this design shows a simple configuration and a miniaturized size. The filter example is designed at 4.2 GHz with a 3 dB fractional bandwidth of 55%, which can meet the low band of 3.1–5.1 GHz of the DS-UWB specification with a slight tuning of the passband. Moreover, the filter design procedure is flexible and can be designed at other frequency ranges when using the desired substrate. Therefore, because of its simple topology, miniaturized and compact size, and good performance, the designed filter is very useful for modern wideband wireless communication systems.

Table 1. Comparison of filter performances of the proposed filter with the previous published works.

	Ref. [11]	Ref. [12]	Ref. [13]	Ref. [14]	Ref. [15]	This Work
Center frequency (GHz)	2.3	3	2	2.3	1	4.2
$|S_{11}|$ (dB)	13	11.7	20	>13	15	15
$|S_{21}|$ (dB)	0.35	2.1	0.57	0.35	1	1.2
3 dB FBW (%)	80	107	100	80	123	55
Circuit Size (λg × λg)	0.12 × 0.22	0.89 × 0.46	No description	0.53 × 0.43	0.17 × 0.14	0.3 × 0.1
Wide stopband	No	No	Yes	No	Yes	Yes
Defected ground	No	Yes	No	No	No	No

4. Conclusions

This paper presented a simple method to design a miniaturized wideband bandpass filter with suppression of the third harmonic. The filter is formed by simply using only two quarter-wavelength stepped-impedance resonators (SIRs). For the wideband design, the resonant modes of the two quarter-wavelength SIRs are chosen first by selecting the impedance ratio (K), and length ratio (α), to the lower band edge of the designed wideband. The I/O ports are then directly taped using a similar 0° feeding structure at the desired position of the two opposite direction quarter-wavelength SIRs. With a suitable arrangement of the two SIRs, the wideband response can be formed. A designed filter with a 3 dB fractional bandwidth of 55% was presented and fabricated in this study to verify the design concept. The measured results have a center frequency of 4.2 GHz with a low insertion loss of 1.5 dB, a return loss greater than 15 dB, and a stopband from 7.5 to 12 GHz with an attenuation of 15 dB. A transmission zero appears at 8.3 GHz to obtain an acceptable band selectivity. The average group delay is as low as it is around 0.1–0.75 ns. Moreover, the filter has a miniature size due to this simple design topology. The simulated results and the measured results are in good agreement. Based on this design concept, further works will be performed on the design of the dual wideband filter and diplexer in a miniature size.

Author Contributions: Conceptualization, M.-H.W.; methodology, L.L. and P.Z.; software, L.L.; validation, L.L. and P.Z.; formal analysis, L.L. and P.Z.; investigation, M.-H.W.; resources, R.-Y.Y.; data curation, C.-Y.T.; writing—original draft preparation, L.L. and M.-H.W.; writing—review and editing, L.L. and R.-Y.Y.; visualization, L.L. and P.Z.; supervision, R.-Y.Y.; project administration, R.-Y.Y.; funding acquisition, M.-H.W.

Funding: This work was supported by the Putian University's Initiation Fee Project for Importing Talents for Scientific Research (2019001) and (2019003).

Acknowledgments: The authors acknowledge Hong-Zheng Lai and Shih-Kun Liu for the help with sample measurement.

Conflicts of Interest: The authors declare no conflict of interest.

References

1. *Revision of Part 15 of the Commission's Rules Regarding Ultra-Wideband Transmission Systems*; ET-Docket 98-153, FCC02-48; Federal Communication Commission: Washington, DC, USA, 2002.

2. Hong, J.S.; Lancaster, M.J. *Microstrip Filters for RF/Microwave Applications*; John and Wiley Sons: New York, NY, USA, 2001.

3.   Chang, D.C.; Hsue, C.W. Wide-band equal-ripple filters in nonuniform transmission lines. *IEEE Trans. Microw. Theory Tech.* **2002**, *50*, 1114–1119. [CrossRef]

4.   Hung, C.Y.; Weng, M.H.; Yang, R.Y.; Su, Y.K. Design of the parallel coupled wideband bandpass filters with very high selectivity and wide stopband. *IEEE Microw. Wirel. Compon. Lett.* **2007**, *17*, 510–5127. [CrossRef]

5.   Zhu, L.; Sun, S.; Menzel, W. Ultra-wideband (UWB) bandpass filters using multiple-mode resonator. *IEEE Microw. Wirel. Compon. Lett.* **2005**, *15*, 796–798.

6.   Li, R.; Zhu, L. Compact UWB bandpass filter using stub-loaded multiple-mode resonator. *IEEE Microw. Wirel. Compon. Lett.* **2007**, *17*, 40–42. [CrossRef]

7.   Killamsetty, V.K.; Mukherjee, B. Compact wideband bandpass filter for TETRA band applications. *IEEE Microw. Wirel. Compon. Lett.* **2017**, *27*, 630–632. [CrossRef]

8.   Wang, H.; Zhu, L.; Menzel, W. Ultra-wideband bandpass filter with hybrid microstrip/CPW structure. *IEEE Microw. Wirel. Compon. Lett.* **2005**, *15*, 844–846. [CrossRef]

9.   Ye, C.S.; Su, Y.K.; Weng, M.H.; Kuan, H.; Syu, J.J. Design of a compact CPW bandpass filter used for UWB application. *Microw. Opt. Tech. Lett.* **2009**, *51*, 298–300. [CrossRef]

10. Chang, Y.C.; Kao, C.H.; Weng, M.H.; Yang, R.Y. Design of compact wideband bandpass filter using with low loss, high selectivity and wide stopband. *IEEE Microw. Wirel. Compon. Lett.* **2008**, *18*, 770–772. [CrossRef]

11. Choudhary, D.K.; Chaudhary, R.K. A compact via-less metamaterial wideband bandpass filter using split circular rings and rectangular stub. *Prog. Electromagn. Res. Lett.* **2018**, *72*, 99–106. [CrossRef]

12. Ji, X.C.; Ji, W.S.; Feng, L.Y.; Tong, Y.Y.; Zhang, Z.Y. Design of a novel multi-layer wideband bandpass filter with a notched band. *Prog. Electromagn. Res. Lett.* **2019**, *82*, 9–16. [CrossRef]

13. Li, Z.; Wu, K.L. Direct synthesis and design of wideband bandpass filter with composite series and shunt resonators. *IEEE Trans. Microw. Theory Technol.* **2017**, *65*, 3789–3800. [CrossRef]

14. Li, Y.; Choi, W.W.; Tam, K.W.; Zhu, L. Novel wideband bandpass filter with dual notched bands using stub-loaded resonators. *IEEE Microw. Wirel. Compon. Lett.* **2017**, *27*, 25–27.

15. Gao, X.; Wen, F.; Che, W. Compact ultra-wideband bandpass filter with improved upper stopband. *IEEE Microw. Wirel. Compon. Lett.* **2007**, *17*, 643–645.

16. Hameed, M.; Xiao, G.; Najam, A.I.; Qiu, L.; Hameed, T. Quadruple-mode wideband bandpass filter with improved out-of-band rejection. *Electronics* **2019**, *8*, 300. [CrossRef]

17. Makimoto, M.; Yamashita, S. Bandpass filters using parallel coupled stripline stepped impedance resonators. *IEEE Trans. Microw. Theory Tech.* **1980**, *28*, 1413–1417. [CrossRef]

18. Ye, C.S.; Su, Y.K.; Weng, M.H.; Hung, C.Y.; Tang, R.Y. Design of the compact parallel-coupled lines wideband bandpass filters using image parameter method. *Prog. Electromagn. Res.* **2010**, *100*, 153–173. [CrossRef]

19. IE3D Simulator; Zeland Software, Inc.: Fremont, CA, USA, 2002.

20. Tsai, C.M.; Lee, S.Y.; Tsai, C.C. Performance of a planar filter using a 0° feed structure. *IEEE Trans. Microw. Theory Tech.* **2002**, *50*, 2362–2367. [CrossRef]

 *electronics*

Article

Additive Manufacturing of Monolithic Microwave Dielectric Ceramic Filters via Digital Light Processing

Qingrong Liu [1], Mingbo Qiu [1], Lida Shen [1,*], Chen Jiao [1], Yun Ye [1], Deqiao Xie [1], Changjiang Wang [2], Meng Xiao [3] and Jianfeng Zhao [1]

[1] College of Mechanical and Electrical Engineering, Nanjing University of Aeronautics and Astronautics, Nanjing 210016, China; liuqr@nuaa.edu.cn (Q.L.); qiumingbo@nuaa.edu.cn (M.Q.); jiaochen@nuaa.edu.cn (C.J.); yeyun@nuaa.edu.cn (Y.Y.); dqxie@nuaa.edu.cn (D.X.); zhaojf@nuaa.edu.cn (J.Z.)

[2] Department of Engineering and Design, University of Sussex, Sussex House, Brighton BN1 9RH, UK; C.J.Wang@sussex.ac.uk

[3] Nanjing Institution of Advanced Laser Technology, Nanjing 210038, China; xiaomeng@siom.ac.cn

* Correspondence: ldshen@nuaa.edu.cn; Tel.: +86-189-5189-2566

Received: 26 August 2019; Accepted: 18 September 2019; Published: 20 September 2019

Abstract: Microwave dielectric ceramics are employed in filters as electromagnetic wave propagation media. Based on additive manufacturing (AM) techniques, microwave dielectric ceramic filters with complex and precise structures can be fabricated to satisfy filtering requirements. Digital light processing (DLP) is a promising AM technique that is capable of producing filters with high accuracy and efficiency. In this paper, monolithic filters made from Al_2O_3 and TiO_2, with a molar ratio of 9:1 $(0.9\ Al_2O_3\text{-}0.1\ TiO_2)$, were fabricated by DLP. The difference in the dielectric properties between the as-sintered and post-annealed samples at different temperatures was studied. The experimental results showed that when sintered at 1550 °C for 2 h and post annealed at 1000 °C for 5 h, $0.9\ Al_2O_3\text{-}0.1$ TiO_2 exhibited excellent dielectric properties: ε_r = 12.4, $Q \times f$ = 111,000 GHz, τ_f = +1.2 ppm/°C. After comparing the measured results with the simulated ones in the passband from 6.5 to 9 GHz, it was concluded that the insertion loss (IL) and return loss (RL) of the filter meet the design requirements.

Keywords: microwave dielectric ceramics; filter; additive manufacturing; digital light processing; post annealing; dielectric properties

---

1. Introduction

In recent years, microwave dielectric ceramics have shown great advantages in the development of miniaturization, weight reduction, and integration of filters [1]. Low dielectric constant microwave dielectric ceramics have fast transmission and response speed, high temperature stability in working environment, high signal transmission quality, low transmission loss, and good frequency selectivity. Owing to these superior qualities, they are widely used in 5G wireless mobile communication, satellite communication, and radar systems.

When the microwave dielectric ceramic is employed on equipment with high communication quality, it is necessary to have a small dielectric constant ε_r, a quality factor Q as large as possible, and a frequency temperature coefficient τ_f of near zero in order to meet the requirements of the application. Actually, the microwave communication device is used at different temperatures. If the resonant frequency of the microwave dielectric material changes greatly with temperature, the carrier signal will drift at different temperatures, thereby affecting the performance of the device. It is required that the resonant frequency of the material should hardly change with temperatures. Mg_2SiO_4, $Mg_4Nb_2O_9$, and Al_2O_3 are widely studied as common low dielectric constant microwave dielectric ceramics, though the $Q \times f$ value of Al_2O_3 is significantly higher than that of Mg_2SiO_4 and $Mg_4Nb_2O_9$ [2]. As a typical low dielectric constant microwave dielectric ceramic, Al_2O_3 possesses the qualities of

low manufacturing cost, high precision and great thermal conductivity (i.e., 24.5 W m^{-1} K^{-1} at room temperature) [3]. Above all, it exhibits outstanding dielectric properties at millimeter wave frequency. Alford reported the microwave dielectric properties of Al_2O_3: ε_r = 10, Q × f = 500,000 GHz, and τ_f = −60 ppm/°C [4]. However, it also has obvious disadvantages. As mentioned above, the τ_f of Al_2O_3 is relatively large and the sintering temperature is quite high (1600 °C to 1700 °C), limiting its industrial application. At the same time, TiO_2 ceramics have a ε_r of 100, a Qf of 48,000 GHz, and a positive large τ_f of + 450 ppm/°C [5]. Therefore, in order to improve the dielectric performance of Al_2O_3 ceramics, a certain amount of TiO_2 is usually added to form Al_2O_3-TiO_2 composite ceramics, allowing a value of τ_f tending to zero to be obtained. Youshihiro et al. have reported the dielectric properties of the Al_2O_3-TiO_2 system, of which τ_f is nearly zero when the molar ratio of Al_2O_3 to TiO_2 is 9:1 [6]. Due to different working temperatures, the thermal conductivity of the Al_2O_3-TiO_2 system is a problem that needs to be considered. Generally, high thermal conductivity of typical ceramics can be correlated to common features like simple crystal structure, low atomic mass, strong interatomic forces, high atomic packing density, comparable atomic weight differences among the components in the composition, etc. Moreover, the thermal conductivity has a reciprocal relation with the square root of mean atomic mass in the material [7]. It is worth noting that the elements in 0.9 Al_2O_3-0.1 TiO_2 system possess relatively low atomic weight. The melting temperature of 0.9 Al_2O_3-0.1 TiO_2 is around 1900 °C, which is an indirect indicator of strong interatomic bonding, a prerequisite for high thermal conductivity.

Nowadays, microwave and radio frequency passive devices realized by additive manufacturing (AM) have attracted an increasing amount of attention [8–10]. Different materials can be manufactured via AM, including plastics, metals, and ceramics. The latter has been widely used in filters as dielectric resonators. For components made entirely of ceramics, their surface metalization is achieved by employing electroless nickel/copper plating, thereby avoiding the radiation of the electromagnetic field [11]. A remarkable advantage of AM compared to traditional computer numerical control (CNC) milling technology, is that the processed object is accumulated layer by layer, and the processing cost is theoretically only related to the volume of the object (i.e., the amount of raw materials) and does not consider the complexity of its geometric structure. Therefore, in terms of complex geometric structures, the cost of AM is lower than that of traditional CNC milling technology, making AM technology particularly suitable for manufacturing complex and precise mechanical structures. Typically, molding accuracy and efficiency are unable to be simultaneously achieved, particularly for AM methods. The technique of point-by-point scanning molding using 3D Gel Printing (3DGP) and stereolithography (SLA) methods requires a significant amount of shaping time [12]. Digital light processing (DLP) based on the surface exposure AM method can theoretically avoid this shortcoming and does not require scanning path planning [13,14]. Compared with other AM methods, DLP has the characteristics of high precision, small internal size, and fast forming speed. Our team has successfully fabricated bone scaffolds using bio-ceramic material hydroxyapatite via DLP AM technology, which demonstrated the feasibility of the DLP method in manufacturing ceramic materials [15]. This study utilizes DLP AM technology to fabricate microwave dielectric ceramic filters to significantly improve the forming efficiency while also ensuring accuracy.

In this paper, a ceramic suspension with suitable viscosity was prepared using photopolymer, Al_2O_3, and TiO_2 powders. The designed monolithic microwave dielectric ceramic filter was fabricated by DLP technology. The filter sample was then obtained after post treatment processes, including drying, debinding, sintering, post annealing, and metalization. The effects of sintering temperature and annealing process on the dielectric properties of 0.9 Al_2O_3-0.1 TiO_2 ceramics were investigated based on the densification, X-ray diffraction (XRD), and microstructure of the samples. The S-parameters and group delay obtained from the actual test were compared with the simulation results, indicating that the manufactured filter meets the design requirements. The research results show that it is feasible to make sophisticated microwave dielectric ceramic filters using DLP AM technology.

2. Materials and Methods

2.1. Preparation of the Ceramic Suspension

Figure 1 shows the complete preparation of the microwave dielectric ceramic filter. In this study, a micro-sized, nano-sized spherical Al_2O_3 powder and micro-sized spherical TiO_2 powder ($Al_2O_3 \geq 99.9$ wt%, $TiO_2 \geq 99.5$ wt%, Chengdu Kewan Intelligent Technology Co., Ltd, Chengdu, China) were used as the starting material with a particle size of 5 mm, 500 nm, and 1 μm, respectively. Table 1 lists the basic physical parameters of the three powders. The Al_2O_3 powder of two particle sizes is uniformly mixed at a 1:1 wt ratio, followed by the TiO_2 powder being added to the mixed Al_2O_3 powder at a molar wt ratio of 1:9. The premixed solution used to prepare the ceramic suspension consisted of two components: photosensitive resin (Shanghai Ai Rui Technology Co., Ltd, Shanghai, China) and sodium polyacrylate dispersant (Hebei Jinghong Chemical Co., Ltd, Hebei, China). The as-prepared powder was then added to the premixed solution to form the ceramic suspension. After manual premixing, the ceramic suspension was then ball-milled for 12 h using zirconia balls. The suspension was then degassed for 30 min using a vacuum mixer. In this way, a ceramic suspension with solid content of 45% could be obtained. In this work, the mixing of the nanopowders and the micropowders was performed in order to create a balance between the viscosity of the ceramic slurry and the density of the sintered body. Previous theoretical and experimental studies have demonstrated that the combination of powders with different particle sizes is an effective method for increasing the volume fraction and reducing the viscosity of the ceramic slurry. Moreover, the density of the sintered body made of powders of different particle sizes is higher than that made of a single particle size powder [16].

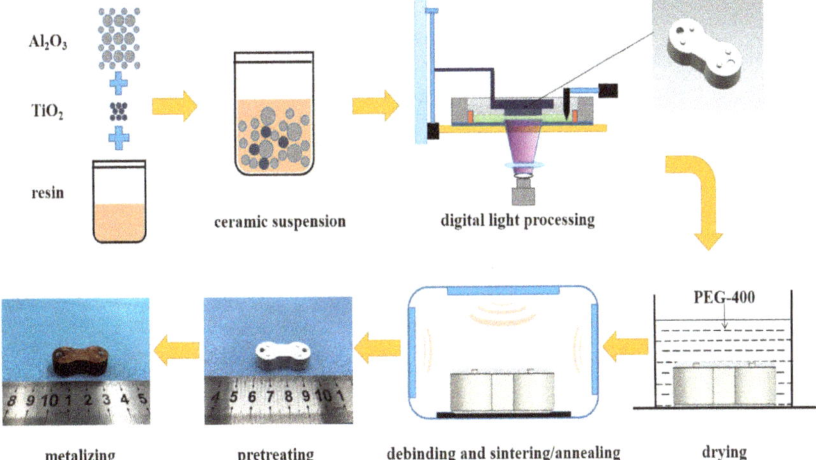

Figure 1. Preparation processes of the microwave dielectric ceramic filter.

Table 1. Characteristic parameters of the raw material.

	AL-1	AL-2	TiO_2
D50 (μm)	5	0.5	2
Density (g·cm^{-3})	3.726	3.742	4.124
Density (g·cm^{-3})	2.136	10.996	13.216

2.2. Design and Simulation of the Microwave Dielectric Ceramic Filter

As depicted in Figure 2a, a two-pole transverse magnetic (TM) mode microwave dielectric ceramic filter, in which the fundamental resonance is the $TM_{01\delta}$ mode, was designed with the following parameters. The center frequency was 8.3 GHz and fractional bandwidth was 1.54%. The height of the filter was 10 mm, whereas the radius of the cylindrical cavity was 7.58 mm. The diameter of the two center posts, acting as resonators, was 3 mm. The thickness of the chamber wall was set to 1 mm, which is convenient for manufacturing and microwave performance testing. Alignment posts were designed to facilitate the placement of Sub-Miniature version A (SMA) connectors for the input and output couplings. After designing the size parameters of the filter, a High Frequency Structure Simulator (HFSS) was utilized for simulation, which is based on the finite element method (FEM). The essential steps are listed below.

(1)   The geometric model of the filter was first created using 3D Modeler according to the size parameters of the design, which are shown in Figure 2b.
(2)   The microwave dielectric material selected in the design was set to have a ε_r of 12.4.
(3)   According to the defined port and boundary conditions, excitation type was a lumped port excitation, boundary type was set to be Perfect E.
(4)   Then the simulation frequency range needed to be defined. The center frequency of the designed filter was 8.3 GHz, and the simulation frequency range was 6.5–9 GHz.
(5)   After the above operations, the simulation step was run.

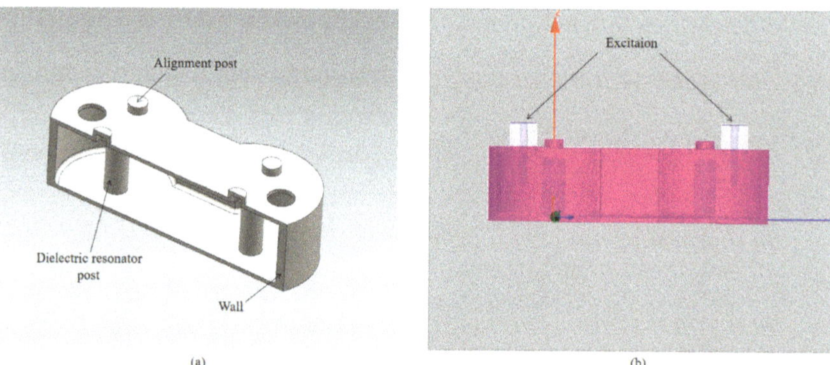

Figure 2. Prototype of the designed microwave dielectric ceramic filter. (**a**) Structure geometry and (**b**) simulation layout.

2.3. Fabrication and Post Treatment of the Filters

Firstly, the 3D model was created using the Unigraphics NX (UG) software (Siemens PLM Software, Plano, TX, USA), and then the Magics software was employed to generate the supporting structure and to slice the parts. The final data was output as a stereolithography (STL) file, which was then imported into the DLP printer. The DLP printing machine was developed by the Nanjing University of Aeronautics and Astronautics. The light source of the DLP printer can emit ultraviolet light at a wavelength of 405 nm. The size of the experimental molding substrate was 60 mm × 80 mm, and its z axis accuracy was 10 µm. As shown in Figure 3, selective scanning of the ceramic suspension by ultraviolet light during the molding process cures each layer pattern. After the first layer is cured, the forming platform moves upward and the ceramic suspension was recoated on the cured surface with a blade. The second layer is then cured using the same process. These steps are repeated until the entire ceramic body is finally obtained.

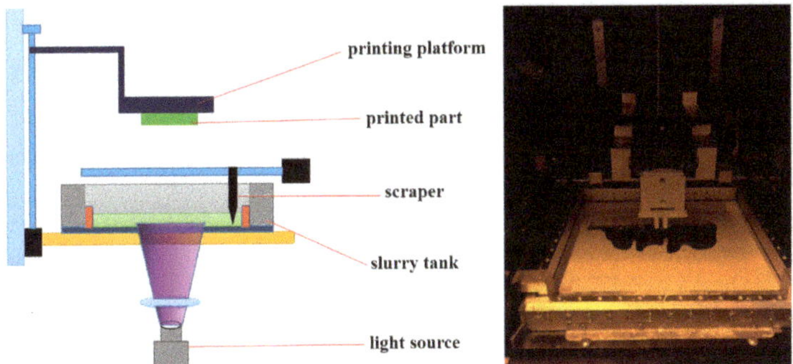

Figure 3. Schematic view of the printing process.

The printed ceramic body then underwent ultrasonic cleaning in alcohol, followed by running tap water. After that, the residual water in the body had to be removed by drying. An innovative method is to use PEG-400. For the PEG-based extraction process, the sample was immersed in PEG-400, which was expected to result in a uniform extraction rate in all directions [17]. With the purpose of removing the polymer binder and achieving densification, the sample then underwent two thermal processes: debinding and sintering. A two-step debinding method was adopted in this work. The first stage under vacuum debinding was to slow down the pyrolysis rate, and a second debinding in the air was conducted to ensure complete removal of residual carbon. The debinding stage is the most time-consuming step since the resin needs to be slowly removed to prevent cracking. As shown in Figure 4, according to the phase diagram of Al_2O_3-TiO_2, the two reacted at 1200 °C to form the Al_2TiO_5 phase. The appearance of Al_2TiO_5 depreciates the properties of Al_2O_3-TiO_2 microwave dielectric ceramics [6]. Accordingly, the annealing process is required to eliminate the adverse effects of Al_2TiO_5. The binder-removed samples were sintered at temperatures of 1450 °C, 1500 °C, 1550 °C, and 1600 °C. They were then kept at the highest temperature for 2 h, and then annealed at 1000 °C for 5 h. The sintered 0.9 Al_2O_3-0.1 TiO_2 ceramic part shrinks, and the shrinkage coefficient of the sample was measured to be approximately 25%. This signifies that the original model must be scaled up accordingly before manufacturing. In order to shield the filter, a 10 μm copper layer was applied to the external surfaces.

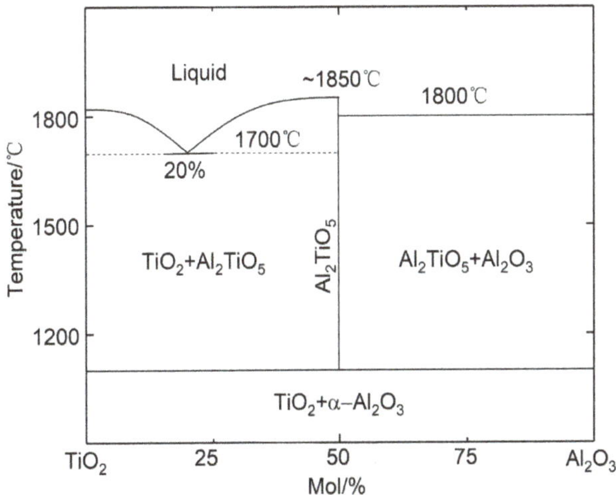

Figure 4. Phase diagram of TiO_2-Al_2O_3.

2.4. Measurements and Characterizations

The microstructure of the sintered sample was characterized by scanning electron microscope (SEM) (S-4800; Hitachi Instruments, Tokyo, Japan). The relative density was determined by the Archimedes' displacement method. The samples at different temperatures were investigated by XRD (DMAX2500PC; Rigaku Corp., Tokyo, Japan) under the following conditions: 10°–80° diffraction angle, CuKα radiation (λ = 1.5406 Å), 40 kV, 100 mA, 0.02° step width, and 5°/min scanning speed. Qf and ε_r were measured by a network analyzer (Agilent 8722E) (Keysight, Santa Rosa, CA, USA) using a pair of parallel conducting Ag and Cu plates in the TE_{011} mode of modified Hakki and Coleman's resonator method [18,19]. The Agilent 8722ET network analyzer and incubator were used to measure τ_f at the temperature range of 20 °C–80 °C. The calculation formula is:

$$\tau_f = \frac{1}{f_{20}} \cdot \frac{f_{80} - f_{20}}{80 - 20} \tag{1}$$

in which f_{80} and f_{20} are the resonant frequencies at 80 °C and 20 °C, respectively. The designed microwave dielectric ceramic filter was simulated by HFSS, and its insertion loss, return loss, and group delay were obtained. As shown in Figure 5, two SMA connectors were mounted on the fabricated filter structure to enable testing of their microwave performance with the network analyzer (Keysight N5247A) (Keysight, Santa Rosa, CA, USA).

Figure 5. Schematic diagram of mounting position of two SMA connectors on the filter.

3. Results

3.1. The Results of the Dielectric Properties

Figure 6 shows the changing trend in the dielectric properties of the 0.9 Al_2O_3-0.1 TiO_2 system, and Table 2 lists the accurate values of the 0.9 Al_2O_3-0.1 TiO_2 system before and after annealing at different sintering temperatures. As shown in Figure 6a, with the increase in sintering temperature, the Qf value of the 0.9 Al_2O_3-0.1 TiO_2 ceramics first increases and then decreases, reaching a maximum at 1550 °C. However, the post-annealed sample had an increase of about 30,000 GHz in Qf value compared to the as-sintered sample at the same temperature. From the point of view of ceramic technology, as long as there exists consistent structure, high density, and uniform grain growth, impurities and defects are reduced. As such, the dielectric loss tanδ can be reduced, thus resulting in an improved Q value. In the Discussion section, dielectric properties will be analyzed based on the relative density, microstructure, and crystal structure of the sample.

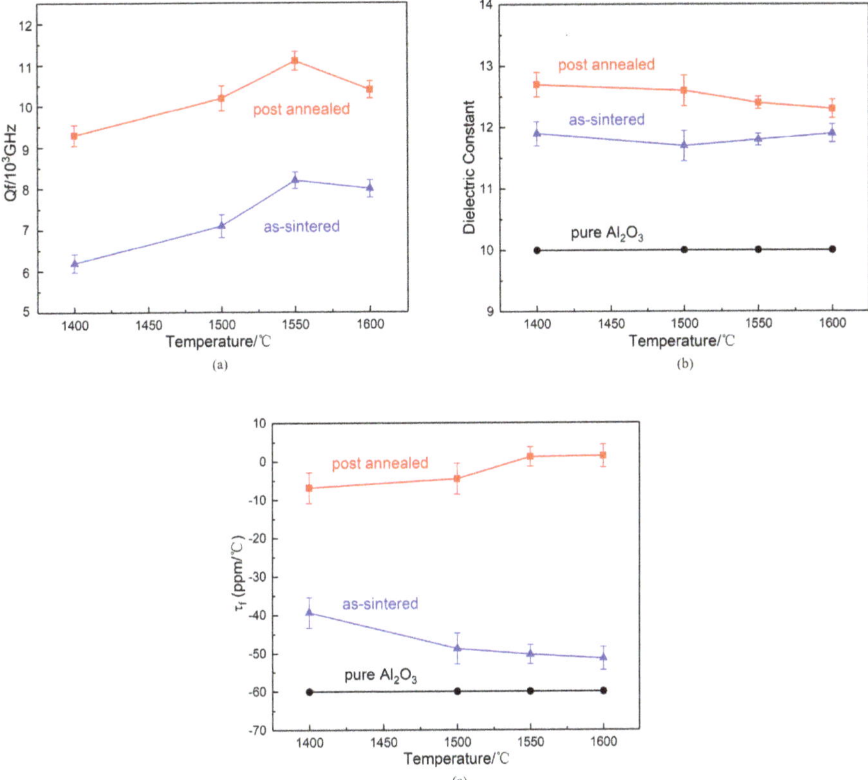

Figure 6. (a) Qf, (b) ε_r, and (c) τ_f, of 0.9 Al_2O_3-0.1 TiO_2 ceramics sintered at different temperatures for 2 h and annealed at 1000 °C for 5 h. The results of pure Al_2O_3 are also plotted in (**b**,**c**) for comparison [3].

Table 2. Microwave dielectric properties of 0.9 Al_2O_3-0.1 TiO_2 ceramics.

Heat Treatment	$Q \cdot f$ (GHz)	ε_r	τ_f (ppm/°C)
1400 °C as-sintered	62,000	11.9	−39.3
1400 °C post annealed	93,000	12.7	−6.8
1500 °C as-sintered	71,000	11.7	−48.7
1500 °C post annealed	102,000	12.6	−4.5
1550 °C as-sintered	82,000	11.8	−50.2
1550 °C post annealed	111,000	12.4	+1.2
1600 °C as-sintered	80,000	11.9	−51.3
1600 °C post annealed	104,000	12.3	+1.4

Figure 6b shows that the ε_r of the 0.9 Al_2O_3-0.1 TiO_2 system is larger than that of the pure Al_2O_3, while the ε_r of the post-annealed sample is increased compared with the as-sintered sample. It can be seen from Table 2 that the dielectric constant of the 0.9 Al_2O_3-0.1 TiO_2 system after annealing at different temperatures is essentially unchanged, all are around 12.5.

Figure 6c shows the variation of the τ_f of the 0.9 Al_2O_3-0.1 TiO_2 system at different temperatures. With the increase in temperature, the τ_f of pure Al_2O_3 remains basically unchanged, which is approximately equal to −60 ppm/°C. The sample after annealing has a larger improvement than the as-sintered in the τ_f, and the value of τ_f differs by more than 30 ppm/°C. After annealing at 1550 °C, the τ_f of the sample is +1.2 ppm/°C, which is the closest to zero.

3.2. The Results of S Parameters and Group Delay of the Filter

The aforementioned dielectric properties of 0.9 Al_2O_3-0.1 TiO_2 at different temperatures indicates that when sintered at 1550 °C for 2 h and annealed at 1000 °C for 5 h, the composite material exhibits the best performance. The DLP-manufactured filter was treated by this process.

Figure 7a shows the differences of insertion loss (IL) and return loss (RL) of the microwave dielectric ceramic filter between the actual measurement and electromagnetic simulation. The center frequency of the filter was set to be 8.3 GHz in the electromagnetic simulation whereas the measured center frequency deviates from the center frequency by 0.05 GHz, that is the Δf = 0.6%. As shown in Figure 7a, the IL in the passband of 8.24–8.37 GHz in the actual measurement is 1.16–2.52 dB, which is 1.15–2.49 dB higher than the simulated one. The measured S11 curve shows that the RL is greater than 30 dB in the passband, meeting the performance specifications required for this type of filter. The passband group delay of the filter is plotted in Figure 7b. The actual measurement results show that the group delay in the 3 dB bandwidth (8.24–8.37 GHz) of the passband is 0.87–1.48 ns. Figure 7c indicates that it has good isolation and output RL in actual measurement.

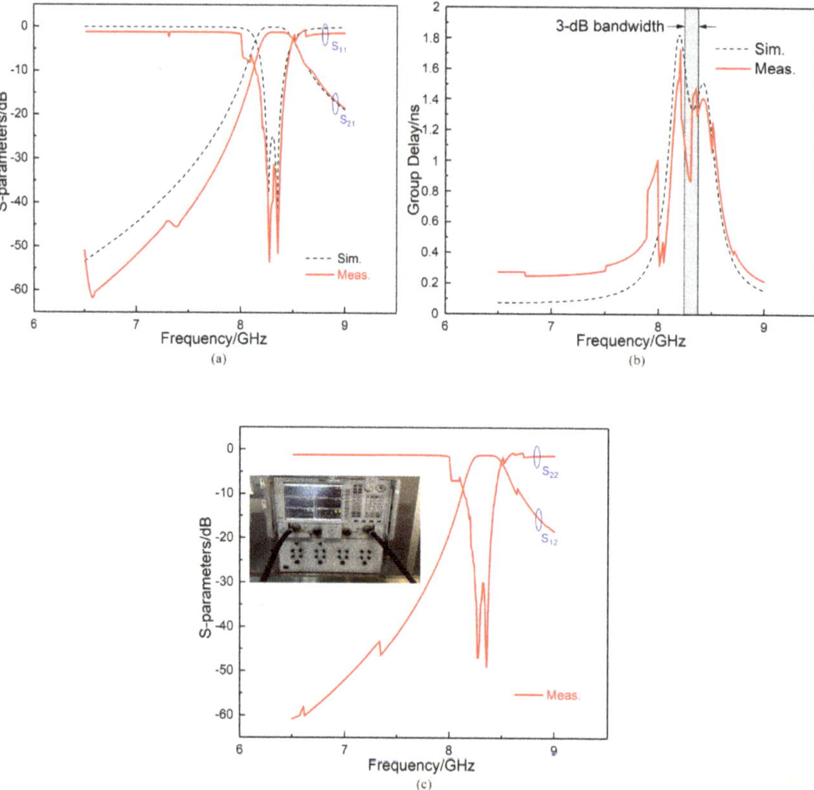

Figure 7. (a) Wideband responses of the S11 and S21 parameters. (b) Passband group delay. (c) Measured wideband responses of the S12 and S22 parameters. The inset shows the photograph of the network analyzer.

4. Discussion

4.1. Discussion of the Dielectric Properties

The quality of the microstructure has a great influence on the dielectric properties of microwave dielectric ceramics. Figure 8(a1) shows that when the sintering temperature is 1400 °C, the as-sintered sample has different grain sizes and no obvious grain boundaries. However, as is shown in Figure 8(b1), the grains of the annealed sample are improved and the surface pores are reduced. When the sintering temperature rises, the grain size tends to be uniform and the surface pores gradually reduce in number and size. When the sintering temperature rises to 1600 °C, some grains are excessively grown due to over burning. It can be observed in Figure 8(a3,b3) that when the sample is sintered at 1550 °C for 2 h and annealed at 1000 °C for 5 h, the surface is regular, the grain distribution is uniform, the pores on the surface are small, and the grain boundaries are more obvious. This further verified that the 0.9 Al_2O_3-0.1 TiO_2 system shown in Figure 6 has the best dielectric properties when sintered at 1550 °C for 2 h and post annealed at 1000 °C for 5 h.

The Qf value of 0.9 Al_2O_3-0.1 TiO_2 first increases and then decreases with the increase in the sintering temperature, which is the result of increasing density and the appearance of secondary phase Al_2TiO_5. The Qf value of TiO_2 is 500,000 Hz, which is one order of magnitude lower than that of Al_2O_3 (Qf = 5,000,000 Hz). The 0.9 Al_2O_3-0.1 TiO_2 system can be regarded as adding TiO_2 to the Al_2O_3 system. Firstly, with the addition of TiO_2, the Qf value of the system must be lower than the Qf value of Al_2O_3. As shown in Figure 9, as the sintering temperature gradually increased, the relative density of the sample becomes increasingly higher, reaching a maximum at 1550 °C. Owing to the densification of the material, the tighter the ion binding is, the more difficult the ion movement is. As such, it is difficult for ion relaxation polarization to occur. Overall, there is no polarization loss, except for the electronic and ionic elastic displacement polarization. In addition, when the material is denser, there are fewer pores, thus resulting in the smaller ionization loss caused by gas ionization in the pores. Combining the above two factors, the Qf value first increases. However, when the sintering temperature reaches 1600 °C, some individual particles grow abnormally because of over burning, causing structural defects, which is the main reason for the decrease in the Qf value. Under the joint influence of these factors, the Qf value of the 0.9 Al_2O_3-0.1 TiO_2 system first increases and then decreases with the increase of sintering temperature.

However, the Qf value of the as-sintered sample is lower than that of the annealed one at the same temperature. The reason for this is that the formation of the high loss phase Al_2TiO_5 reduces the Qf value of the system. It can be observed in Figure 10 that in addition to the diffraction peaks of the two phases of Al_2O_3 and TiO_2, the diffraction peaks of Al_2TiO_5 appear at different sintering temperatures, indicating that Al_2O_3 reacts with TiO_2 to form Al_2TiO_5 during sintering. The sample annealed at 1000 °C for 5 h showed only two phases of Al_2O_3 and TiO_2 in the XRD pattern, signifying that the Al_2TiO_5 phase was decomposed by annealing treatment. According to Figure 4, when the 0.9 Al_2O_3-0.1 TiO_2 system was sintered at 1450 °C, 1500 °C, 1550 °C, and 1600 °C for 2 h, and annealed at 1000 °C for 5 h, the composition of the phases was roughly the same.

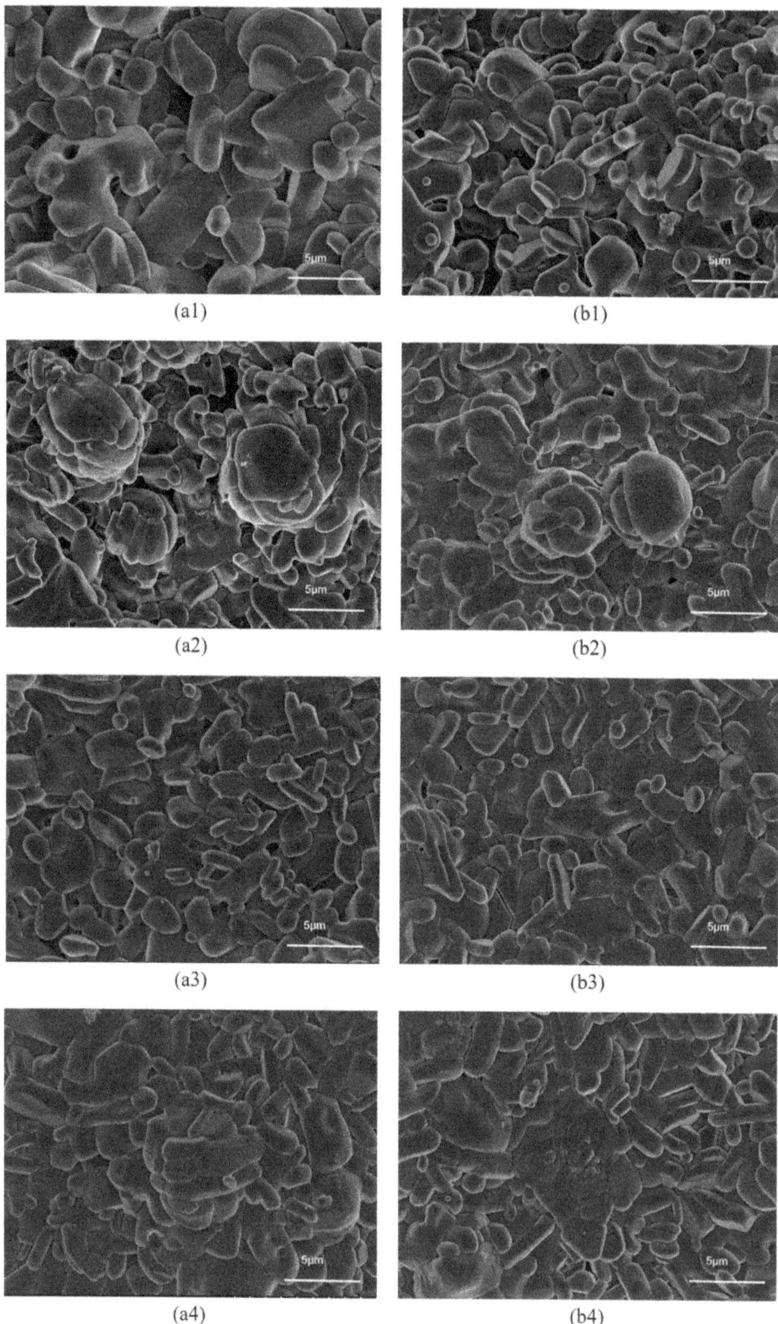

Figure 8. SEM images of 0.9 Al_2O_3-0.1 TiO_2 ceramics sintered at (**a1**) 1400 °C, (**a2**) 1500 °C, (**a3**) 1550 °C, (**a4**) 1600 °C. (**b1**–**b4**) are post-annealed samples of (**a1**–**a4**), respectively.

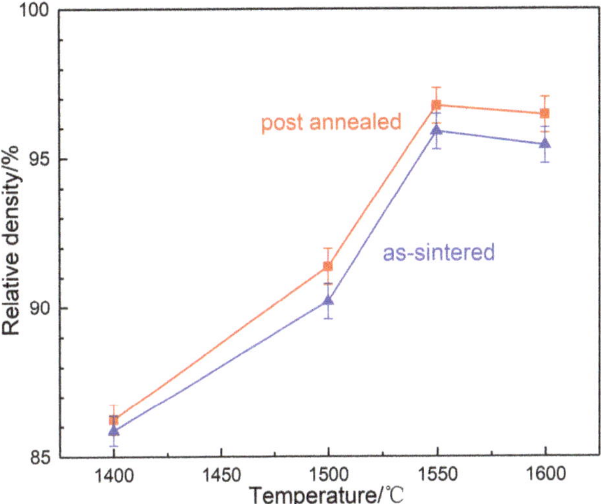

Figure 9. Relative density of 0.9 Al$_2$O$_3$-0.1 TiO$_2$ ceramics sintered at different sintering temperatures and annealed at 1000 °C for 5 h.

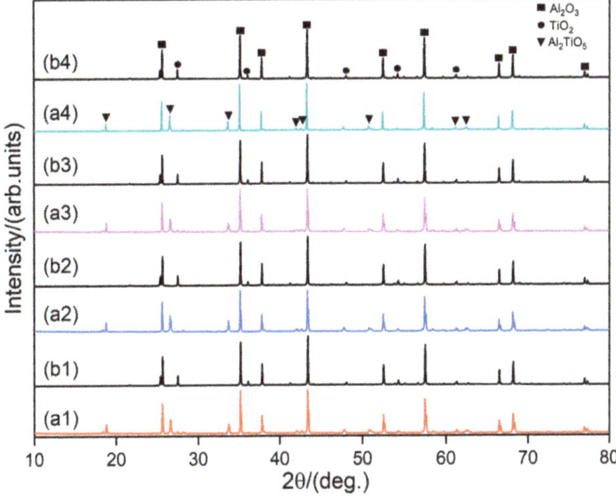

Figure 10. XRD patterns of 0.9 Al$_2$O$_3$-0.1 TiO$_2$ samples sintered at (**a1**) 1400 °C, (**a2**) 1500 °C, (**a3**) 1550 °C, and (**a4**) 1600 °C. (**b1**–**b4**) are post-annealed samples of (**a1**–**a4**), respectively.

Since $\varepsilon_r = 10$ of Al$_2$O$_3$ and $\varepsilon_r = 100$ of TiO$_2$, the ε_r of the system is bound to increase with the addition of TiO$_2$. As shown in Figure 4, when the sintering temperature is higher than 1200 °C and annealing is not performed, the secondary phase Al$_2$TiO$_5$ is formed in the 0.9 Al$_2$O$_3$-0.1 TiO$_2$ system. However, the ε_r value of the Al$_2$TiO$_5$ is 3.4, which is lower than that of the main phase. Annealing at 1000 °C for 5 h results in the decomposition of Al$_2$TiO$_5$ with low ε_r and the precipitation of TiO$_2$ with high ε_r, so the ε_r of the annealed sample is increased compared to the as-sintered one.

According to microscopic analysis, the ε_r is a macroscopic physical quantity that comprehensively reflects the polarization behavior of the dielectric. When the polarizing ability of the dielectric under the electric field is stronger, the ε_r is larger. By Clausius equation

$$P = N\alpha Ei \tag{2}$$

where P: macroscopic polarizability, α: microscopic polarizability, N: number of molecules per unit volume of dielectric, Ei: effective electric field, and E: average macroscopic electric field. So:

$$\varepsilon_r = 1 + \frac{N\alpha}{\varepsilon_0} \bullet \frac{Ei}{E} \tag{3}$$

It can be estimated from Formula (3) that there are three ways to increase ε_r. The first method is to increase the value of N, that is, to increase the density of the dielectric materials. The second method is to adopt dielectric materials composed of plasmids that have a large polarization α. The last method is by selecting material that have a large effective electric field Ei. The ε_r of the Al_2O_3 system mainly depends on the ion polarization, which is predominantly determined by the polarization mode of the lattice vibration. According to RD Shannon [20], the polarizability of Al^{3+} ions is 0.78×10^{-40} $C \cdot m^2 \cdot V^{-1}$, and the polarization of Ti^{4+} ions is 2.94×10^{-40} $C \cdot m^2 \cdot V^{-1}$, which further proves that the introduction of Ti^{4+} ions will increase the ε_r of the $0.9\ Al_2O_3$-$0.1\ TiO_2$ system.

The ε_r of the material satisfies the formula:

$$\frac{1}{\varepsilon_r} = \frac{1}{\varepsilon_g} + \frac{1}{R\varepsilon_{gb}}, \tag{4}$$

where ε_g and ε_{gb} are the dielectric constants of the grain and the grain boundary layer, respectively, and R is the ratio of the thickness of the grain to grain boundary layer (Dg/Dgb). It can be determined from Figure 8 that as the sintering temperature increases, the pores decrease and the unit cell volume increases, which is beneficial to the intracellular ion vibration. Further, the polarization is enhanced, resulting in an increase in the dielectric constant ε_g of the grain and the dielectric constant ε_{gb} of the grain boundary layer. During the densification of the microstructure, the crystal grains grow uniformly, and the grain boundary layers become thinner, causing R to increase. According to Formula (4), ε_r is thereby increased.

According to the Lichnetecker Law, the τ_f of Al_2O_3 is negative, -60 ppm/°C, while the τ_f of TiO_2 is positive, approximately 450 ppm/°C [21]. With the addition of TiO_2, the τ_f of the system drifts positively, and $\tau_f \approx 0$ when annealed at 1550 °C. The τ_f value of pure Al_2O_3 ceramics is not affected by the sintering temperature. The τ_f of the as-sintered $0.9\ Al_2O_3$-$0.1\ TiO_2$ ceramics is about -45 ppm/°C, which is attributable to the reaction of TiO_2 and Al_2O_3 at high temperature to form the second phase Al_2TiO_5. As a result of the reduction of TiO_2 content in the system, the negative τ_f of Al_2O_3 ceramics is not corrected too much. When adopting post-annealment, the Al_2TiO_5 phase decomposes and the amount of TiO_2 increases. The τ_f value of the $0.9\ Al_2O_3$-$0.1\ TiO_2$ system changes from negative to positive. As shown in Table 2, the τ_f value of 1.2 ppm/°C was obtained after being annealed at 1550 °C.

4.2. Error Analysis of the Simulated and Measured Results of the Microwave Dielectric Ceramic Filter

Figure 7 presents the measured results compared with the simulation results. The measured center frequency was shifted backward by 0.05 GHz, which is caused by the uncertainty of the shrinkage of the ceramic sample after debinding and sintering. According to the previous measurements, after the body was sintered, the linear shrinkage was about 25%, so the model is enlarged to 1.34 times of the original size. However, the size of the filter will deviate slightly after sintering. For example, although the actual measurement shows that the wall thickness is 1.15 mm, the wall thickness of the sintered filter is expected to be 1 mm. For such reasons, the results of the measurement may deviate from the simulation results. Compared with the simulation results, the actual measured IL and RL are a little large. This is attributable to the fact that, in addition to the manufacturing errors mentioned above, the influence of the layer thickness and quality of the metal layer on the measurement is also considered. It should be noted that this was not taken into account during the electromagnetic simulation process.

5. Conclusions

In this paper, DLP AM technology and microwave dielectric ceramic material were utilized to fabricate a monolithic microwave dielectric ceramic filter with a complex and precious structure. When sintered at 1550 °C for 2 h and post annealed at 1000 °C for 5 h, the obtained dielectric properties of the 0.9 Al_2O_3-0.1 TiO_2 system were the best: ε_r = 12.4, Q × f = 111,000 GHz, and τ_f = +1.2 ppm/°C. The measured value of the central frequency deviates from that of the simulated central frequency by 0.05 GHz, that is, the Δf = 0.6%. The actual IL value is 1.16–2.52 dB in the passband of 8.24–8.37 GHz and the actual RL value is greater than 30 dB, satisfying the design requirements of this type of filter. This work provides an effective method for manufacturing monolithic microwave dielectric ceramic filters with complex and precious structures, which can be widely used in cellular mobile network systems, communication base stations, television satellite receiving systems, satellite communication, and radar systems, etc. By controlling the composition of the ceramic material and improving the quality of the metalization coating, the dielectric properties of the ceramic material can be further optimized and the applicable frequency band of the filter can be further improved.

Author Contributions: Conceptualization, Q.L. and L.S.; investigation, C.J. and D.X.; methodology, Y.Y. and M.X.; writing—original draft preparation, Q.L.; writing—review and editing, C.W., M.Q. and J.Z.

Funding: This research was funded by the National Key Research and Development Program "Additive Manufacturing and Laser Manufacturing" (Grant No. 2018YFB1105400 and 2018YFB1105801), National Natural Science Foundation of China (Grant No.U1537105 and No.U1532106) and National Key Laboratory of Science and Technology on Helicopter Transmission (Nanjing University of Aeronautics and Astronautics) (Grant No. HTL-A-19G011).

Conflicts of Interest: The authors declare no conflict of interest.

References

1. Raveendran, A.; Sebastian, M.T.; Raman, S. Applications of Microwave Materials: A Review. *J. Electron. Mater.* **2019**, *48*, 2601–2634. [CrossRef]
2. Sebastian, M.T.; Ubic, R.; Jantunen, H. Low-loss dielectric ceramic materials and their properties. *Int. Mater. Rev.* **2015**, *60*, 392–412. [CrossRef]
3. Partridge, G. Inorganic materials V. Ceramic materials possessing high thermal conductivity. *Adv. Mater.* **1992**, *4*, 51–54. [CrossRef]
4. Alford, N.M.; Penn, S.J. Sintered alumina with low dielectric loss. *J. Appl. Phys.* **1996**, *80*, 5895–5898. [CrossRef]
5. Templeton, A.; Wang, X.; Penn, S.J.; Webb, S.J.; Cohen, L.F.; Alford, N.M. Microwave Dielectric Loss of Titanium Oxide. *J. Am. Ceram. Soc.* **2000**, *83*, 95–100. [CrossRef]
6. Ohishi, Y.; Miyauchi, Y.; Ohsato, H.; Kakimoto, K.-I. Controlled Temperature Coefficient of Resonant Frequency of Al_2O_3-TiO_2 Ceramics by Annealing Treatment. *Jpn. J. Appl. Phys.* **2004**, *43*, 749–751. [CrossRef]
7. Roshni, S.B.; Sebastian, M.T.; Surendran, K.P. Can zinc aluminate-titania composite be an alternative for alumina as microelectronic substrate? *Sci. Rep.* **2017**, *7*. [CrossRef] [PubMed]
8. Dimitriadis, A.I.; Debogovic, T.; Favre, M.; Billod, M.; Barloggio, L.; Ansermet, J.-P.; Rijk, D.E. Polymer-Based Additive Manufacturing of High-Performance Waveguide and Antenna Components. *Proc. IEEE* **2017**, *105*, 668–676. [CrossRef]
9. Auria, M.D.; Otter, W.J.; Hazell, J.; Gillatt, B.T.W.; Long-Collins, C.; Ridler, N.M.; Lucyszyn, S. 3-D Printed Metal-Pipe Rectangular Waveguides. *IEEE Trans. Compon. Packag. Manuf. Technol.* **2015**, *5*, 1339–1349. [CrossRef]
10. Liu, B.; Gong, X.; Chappell, W.J. Applications of layer-by-layer polymer stereolithography for three-dimensional high-frequency components. *IEEE Trans. Microwave Theory Tech.* **2004**, *52*, 2567–2575. [CrossRef]
11. Guo, C.; Shang, X.; Lancaster, M.J.; Xu, J. A 3-D Printed Lightweight X-Band Waveguide Filter Based on Spherical Resonators. *IEEE Microwave Wireless Compon. Lett.* **2015**, *25*, 442–444. [CrossRef]
12. Zocca, A.; Colombo, P.; Gomes, C.M.; Günster, J. Additive Manufacturing of Ceramics: Issues, Potentialities, and Opportunities. *J. Am. Ceram. Soc.* **2015**, *98*, 1983–2001. [CrossRef]

13. Zeng, Y.; Yan, Y.; Yan, H.; Liu, C.; Li, P.; Dong, P.; Zhao, Y.; Chen, J. 3D printing of hydroxyapatite scaffolds with good mechanical and biocompatible properties by digital light processing. *J. Mater. Sci.* **2018**, *53*, 6291–6301. [CrossRef]

14. He, R.; Liu, W.; Wu, Z.; An, D.; Huang, M.; Wu, H.; Jiang, Q.; Ji, X.; Wu, S.; Xie, Z. Fabrication of complex-shaped zirconia ceramic parts via a DLP- stereolithography-based 3D printing method. *Ceram. Int.* **2018**, *44*, 3412–3416. [CrossRef]

15. Liu, Z.; Liang, H.; Shi, T.; Xie, D.; Chen, R.; Han, X.; Shen, L.; Wang, C.; Tian, Z. Additive manufacturing of hydroxyapatite bone scaffolds via digital light processing and in vitro compatibility. *Ceram. Int.* **2019**, *45*, 11079–11086. [CrossRef]

16. Zaman, A.A.; Dutcher, C.S. Viscosity of Electrostatically Stabilized Dispersions of Monodispersed, Bimodal, and Trimodal Silica Particles. *J. Am. Ceram. Soc.* **2006**, *89*, 422–430. [CrossRef]

17. Zhou, M.; Liu, W.; Wu, H.; Song, X.; Chen, Y.; Cheng, L.; He, F.; Chen, S.; Wu, S. Preparation of a defect-free alumina cutting tool via additive manufacturing based on stereolithography—Optimization of the drying and debinding processes. *Ceram. Int.* **2016**, *42*, 11598–11602. [CrossRef]

18. Hakki, B.W.; Coleman, P.D. A Dielectric Resonator Method of Measuring Inductive Capacities in the Millimeter Range. *IEEE Trans. Microwave Theory Tech.* **1960**, *8*, 402–410. [CrossRef]

19. Kobayashi, Y.; Katoh, M. Microwave Measurement of Dielectric Properties of Low-Loss Materials by the Dielectric Rod Resonator Method. *IEEE Trans. Microwave Theory Tech.* **1985**, *33*, 586–592. [CrossRef]

20. Shannon, R.D. Dielectric polarizabilities of ions in oxides and fluorides. *J. Appl. Phys.* **1993**, *73*, 348–366. [CrossRef]

21. Kobayashi, Y.; Aoki, T.; Kabe, Y. Influence of Conductor Shields on the Q-Factors of a TEO Dielectric Resonator. *IEEE Trans. Microwave Theory Tech.* **1985**, *33*, 1361–1366. [CrossRef]

Article

A New Low-Voltage Low-Power Dual-Mode VCII-Based SIMO Universal Filter

Leila Safari *, Gianluca Barile, Giuseppe Ferri and Vincenzo Stornelli

Department of Industrial and Information Engineering and Economics, University of L'Aquila, 67100 L'Aquila, Italy
* Correspondence: leilasafari@yahoo.com; Tel.: +39-0862-434439

Received: 24 June 2019; Accepted: 4 July 2019; Published: 9 July 2019

Abstract: In this paper, a new low-voltage low-power dual-mode universal filter is presented. The proposed circuit is implemented using inverting current buffer (I-CB) and second-generation voltage conveyors (VCIIs) as active building blocks and five resistors and three capacitors as passive elements. The circuit is in single-input multiple-output (SIMO) structure and can produce second-order high-pass (HP), band-pass (BP), low-pass (LP), all-pass (AP), and band-stop (BS) transfer functions. The outputs are available as voltage signals at low impedance Z ports of the VCII. The HP, BP, AP, and BS outputs are also produced in the form of current signals at high impedance X ports of the VCIIs. In addition, the AP and BS outputs are also available in inverting type. The proposed circuit enjoys a dual-mode operation and, based on the application, the input signal can be either current or voltage. It is worth mentioning that the proposed filter does not require any component matching constraint and all sensitivities are low, moreover it can be easily cascadable. The simulation results using 0.18 μm CMOS technology parameters at a supply voltage of ±0.9 V are provided to support the presented theory.

Keywords: current mode; universal filter; VCII; voltage conveyor; SIMO filter

---

1. Introduction

Filters are among the most widely used circuits in different areas such as instrumentation, communication, control, and signal processing systems [1–5]. Traditionally, filters were designed in voltage mode using active devices such as operational amplifiers (Op-Amp). However, due to the reduced supply voltage in advanced CMOS technologies and the ever-increasing demand on low-power circuits, the design of high-performance Op-Amp-based filters has become very difficult. The inherent low-voltage nature of current-mode signal processing has motivated a wide investigation on current-mode filters with the aim of achieving better results under low supply voltage restrictions. Among the various types of current-mode filters, the multifunction or universal type which is able to produce several transfer functions simultaneously with the same circuit, is good solution for the ever-increasing demand of the market for low-power circuits. That is why, in recent years, numerous current-mode universal filters have appeared in literature [6–12].

In the universal filter category, second-order (biquadratic) filters play an important role in the field of analog signal processing such as the implementation of phase-locked loop (PLL), frequency modulation (FM), stereo demodulation, etc. [1–5]. They can be classified in single-input-multiple-output (SIMO), multiple-input-single-output (MISO), and multiple-input-multiple-output (MIMO) topologies. A deep investigation in the literature reveals that various current-mode active building blocks such as four-terminal floating nullor (FTFN) [6], differential voltage current conveyor (DVCC) [7], current-controlled current conveyor transconductance amplifier (CCCCTA) [7], current feedback operational amplifier (CFOA) [9], current differencing buffered amplifier (CDBA) [10,11], modified current-controlled current differencing transconductance amplifier (MCCCDTA) [12], operational transconductance amplifier (OTA) [13] LT1228 IC [14], fully differential second-generation current conveyor [15], current follower cascaded transconductance amplifier (CFCTA) [16], and voltage differencing differential difference amplifier [17] have been used in the design of multifunction filters. The common feature of the multifunction filters of [6–8,12] is that their output signal is in the form of a current. The reason is that the used current-mode active building blocks lack a low-impedance voltage output port. Therefore, for applications requiring voltage outputs, these circuits require additional voltage buffers resulting in higher power consumption and chip area. There are a few current-mode building blocks such as CDBA, CFOA, and FTFN that have low-impedance voltage output ports. However, the voltage output multifunction filters based on these active building blocks reported in [6,9–11] suffer from serious drawbacks, the most serious of them being their inability to provide low impedance at all voltage output ports and the consequent requirement of additional voltage buffers. For example, the voltage output multifunction FTFN-based filter of [6] requires an extra voltage buffer at its LP output. Similarly, the multifunction CFOA-based filter of [9] requires an extra voltage buffer for HP output. Although the CDBA-based multifunction filter of [9] provides low impedance for all voltage outputs, it has the load and other passive elements connected to output ports, so the output ports must be designed to have high current drive capability. These will result in a complicated internal circuit for the used CDBA. The filter circuits of [14–18] suffer from a high supply voltage requirement.

Recently a new active building block called the second-generation voltage conveyor (VCII) has attracted the attention of researchers [19,20]. It is the dual of well-known second-generation current conveyor (CCII) and, compared to other active building blocks, enjoys a simple internal structure and features more design flexibility. It is characterized by a low-impedance current input port, a high-impedance current output port and a low-impedance voltage output port. Using the low-impedance current input port, current summing/subtracting operations can be performed very easily. In addition, having a low-impedance voltage output port makes this building block highly suitable for voltage output applications. In other words, the low-impedance voltage output port makes the extra voltage buffer unnecessary. More interestingly, VCII internal structure is composed of a current buffer and a voltage buffer. This simple implementation results in low power consumption and low chip area.

Despite these attractive features, up to now, VCII is not used in the implementation of universal filters. Therefore, in this paper, we take the advantages offered by the VCII block to design a voltage output low-power second-order universal filter. The presented filter topology employs one inverting current buffer (I-CB), three double output VCIIs, five resistors, and three capacitors. I-CB is simply a current buffer with gain of −1. The proposed universal filter is designed in SIMO topology and provides all-pass (AP), band-pass (BP), band-stop (BS), low pass (LP), and high-pass (HP) functions. All the outputs are available in voltage form at the low-impedance voltage output ports of the VCII. Interestingly, the HP, BP, and BS outputs are also available in current form at the high-impedance port of the VCII. Moreover, the AP and BS outputs are provided in both the inverting and non-inverting types. The proposed circuit is in dual mode and, based on the application, the input signal can be either a current or a voltage.

The organization of this paper is as follows. In Section 2 the VCII block and its implementation are presented. In Section 3, the proposed filter topology is given. In Section 4, the non-ideal analysis is performed. The simulation results and a comparative table are presented in Section 5. Finally, Section 6 concludes the paper.

2. The VCII Internal Circuit Design

Figure 1a,b shows the schematic diagram and the symbolic representation of a dual-output VCII, respectively. It has a low-impedance current input Y port. Ideally, the impedance at the Y terminal is zero. The current applied to the Y port is transferred to the X_1 and X_2 ports with a current gain of A_{I1} and A_{I2}, respectively. In the ideal case, the values of A_{I1} and A_{I2} are unity. X_1 and X_2 are high-impedance (ideally ∞) current output ports. The voltage produced at the X_1 and X_2 ports is conveyed to the Z_1 and Z_2 ports, respectively. The voltage gain between X_1–Z_1 and X_2–Z_2 is shown by A_{V1} and A_{V2}, respectively, which are equal to unity in the ideal case. The Z_1 and Z_2 ports are low-impedance (ideally zero) voltage output ports. The relation between the port currents and voltages is:

$$v_Y = r_Y i_Y, i_{X1} = A_{I1} i_Y, i_{X2} = A_{I2} i_Y, V_{Z1} = A_{V1} V_{X1}, V_{Z2} = A_{V2} V_{X1}. \tag{1}$$

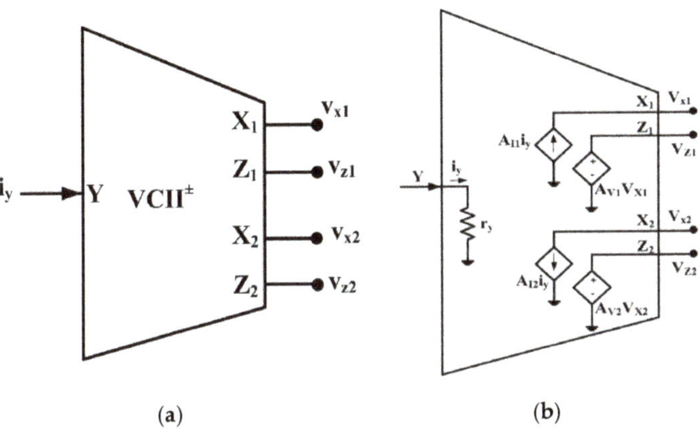

Figure 1. Dual-output second-generation voltage conveyor (VCII), (**a**) symbolic representation and (**b**) internal structure.

As seen from Figure 1, a dual-output VCII is composed of a dual-output current buffer and two voltage buffers. Figure 2 shows the CMOS circuit implementation of the dual-output VCII, representing the dual-output version of the VCII circuit reported in [19]. The current buffers are composed by M_1–M_9, M_{13} transistors together with current sources I_{B1}–I_{B3}, I_{B5}. The input current to the Y terminal is transferred to the X_1 terminal with a current gain of about −1 by an inverting current buffer (I-CB) made of M_1–M_9 transistors and I_{B1}–I_{B3} current sources. The non-inverting current buffer made of M_1–M_6, M_{13} and current sources I_{B1}–I_{B2}, I_{B4} conveys the input current of the Y terminal to the X_2 terminal with a current gain of about +1. The voltage buffers are simply two flipped voltage followers (FVFs) [21] formed by M_{10}–M_{11}, M_{14}–M_{15} transistors and current sources I_{B4}, I_{B6}. The used FVFs provide very low impedance at the Z terminals. The negative-feedback loop established by M_1–M_5 provides very low impedance at the Y terminal. The main feature of the proposed circuit is very low impedance at the Y terminal, which makes this node ideal for a current summing operation. In addition, very low impedance at the Y terminal ensures negligible voltage drop at this node.

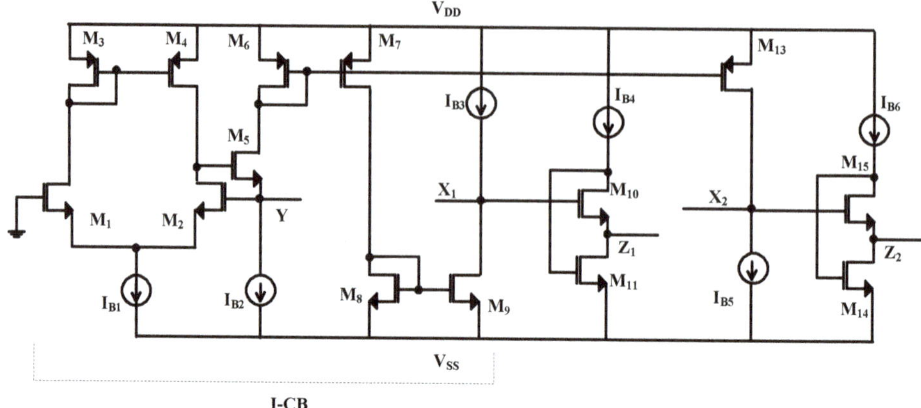

I-CB

Figure 2. CMOS implementation of VCII [19].

3. The Proposed Universal Filter

The proposed VCII-based filter topology is shown in Figure 3. It is based on three dual-output VCIIs and one I-CB as active building blocks and three resistors and three capacitors as passive elements. There is only one floating capacitor. The input signal can be either current or voltage type. Input current is applied to the circuit directly as shown in Figure 3a, while, due to the very low impedance at the Y terminal which is ideally zero, voltage signals can only be applied through a resistor as shown in Figure 3b. The circuit is designed in single-input three-output topology and produces second-order HP, LP, and BP outputs simultaneously. The outputs are available as voltage signals at the low-impedance Z ports of VCII. The HP output can also be provided in the form of current at the high-impedance X_2 port of the VCII. The AP and BS outputs can be simply produced using an additional VCII block and three resistors as shown in Figure 4 in which, as it will be shown, V_{in1}, V_{in2}, and V_{in3} are connected to HP, LP, and BP outputs, respectively. The transfer functions of Figure 3a and b are similar except that, for Figure 3b, the input signal should be replaced with V_{in}/R_{in}. For this similarity, in the following analysis, the transfer functions are only derived for the topology of Figure 3a.

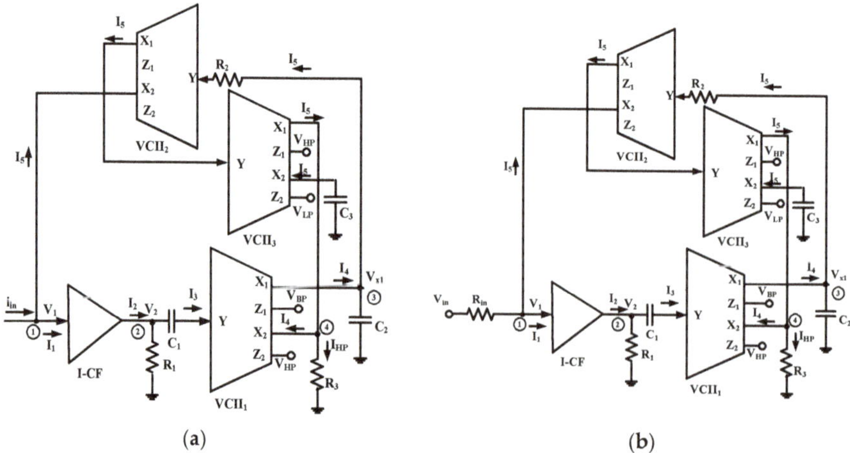

(a)   (b)

Figure 3. The proposed multifunction filter topology with (**a**) current input and (**b**) voltage input.

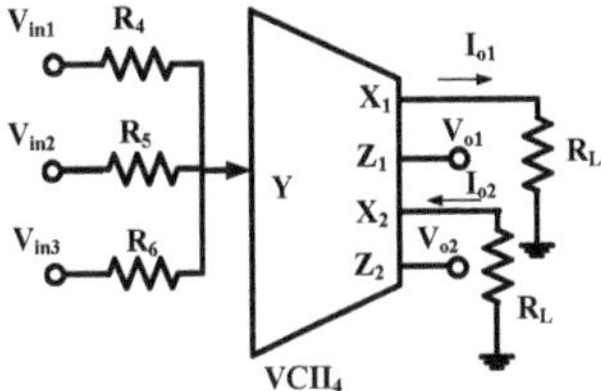

Figure 4. The proposed circuit to produce all-pass (AP) and band-stop (BS) outputs.

Due to the very low impedance at the Y port of the VCII, which is ideally zero, we can assume this port at ground. Therefore, applying Kirchoff's Current Law (KCL) analysis at node 2, gives I_3 as:

$$I_3 = \frac{sC_1R_1}{1 + sC_1R_1}I_2. \tag{2}$$

Using Equation (1), the relationship between I_1–I_2 and I_3–I_4 can be expressed as:

$$I_1 \approx I_2; \; I_3 \approx I_4 \tag{3}$$

By assuming the Y port of VCII$_2$ at ground and performing a KCL analysis at node 3, I_5 is obtained as:

$$I_5 = \frac{1}{1 + sC_2R_2}I_4. \tag{4}$$

At the input node (node 1) we have:

$$i_{in} = I_1 + I_5. \tag{5}$$

Using Equations (3) and (5), I_2 is found as:

$$I_2 = \frac{(1 + sC_1R_1)(1 + sC_2R_2)}{s^2C_1C_2R_1R_2 + s(2C_1R_1 + C_2R_2) + 1}i_{in}. \tag{6}$$

For V_{X1} (the voltage at node 3) we have:

$$V_{X1} = \frac{R_2}{1 + sC_2R_2}I_4. \tag{7}$$

From Equations (2), (3), and (6), I_4 is obtained as:

$$I_4 = \frac{sC_1R_1(1 + sC_2R_2)}{s^2C_1C_2R_1R_2 + s(2C_1R_1 + C_2R_2) + 1}i_{in}. \tag{8}$$

Inserting Equation (8) into Equation (7), gives V_{X1} as:

$$V_{X1} = V_{Z1} = V_{BP} = \frac{sC_1R_1R_2}{s^2C_1C_2R_1R_2 + s(2C_1R_1 + C_2R_2) + 1}i_{in}. \tag{9}$$

Equation (9) represents a second-order BP response. Due to the voltage buffer action between X_1 and Z_1 terminals, the BP response is available at the Z_1 port as a voltage signal.

From Equations (4) and (8), I_5 is found as:

$$I_5 = \frac{sC_1R_1}{s^2C_1C_2R_1R_2 + s(2C_1R_1 + C_2R_2) + 1}i_{in}. \tag{10}$$

By subtracting I_4 (Equation (8)) from I_5 (Equation (10)) the HP output can be provided:

$$I_{HP} = I_5 - I_4 = -\frac{s^2 C_1 R_1 C_2 R_2}{s^2 C_1 C_2 R_1 R_2 + s(2C_1 R_1 + C_2 R_2) + 1} i_{in}. \tag{11}$$

As in Figure 3a, by connecting the current output X_2 ports of $VCII_1$ and $VCII_3$, a HP output in the form of current is produced at node 4. The HP output is also available in the form of voltage at Z_2 ports of $VCII_1$ and $VCII_3$ as:

$$V_{HP} = R_3 I_{HP} = -\frac{s^2 C_1 R_1 C_2 R_2 R_3}{s^2 C_1 C_2 R_1 R_2 + s(2C_1 R_1 + C_2 R_2) + 1} i_{in}. \tag{12}$$

Using Equation (10), the voltage at X_2 and Z_2 ports of $VCII_3$ have a LP transfer function as:

$$V_{X2} \approx V_{Z2} = -\frac{I_5}{sC_3} = V_{LP} = -\frac{\frac{C_1 R_1}{C_3}}{s^2 C_1 C_2 R_1 R_2 + s(2C_1 R_1 + C_2 R_2) + 1} i_{in}. \tag{13}$$

By assuming the Y terminal of $VCII_4$ at ground, the Io_1 and Io_2 outputs are achived as:

$$\left| I_{o1} \right| = \left| I_{o2} \right| = \frac{V_{in1}}{R_4} + \frac{V_{in2}}{R_5} + \frac{V_{in3}}{R_6}. \tag{14}$$

For the AP output, V_{in1}, V_{in2}, and V_{in3} are connected to V_{HP}, V_{BP}, and V_{LP} outputs, respectively. Therefore, by inserting V_{HP} (Equation (12)), V_{BP} (Equation (9)), and V_{LP} (Equation (13)) into Equation (14), the AP transfer function is achieved as:

$$I_{AP} = I_{o1} = I_{o2} = -\frac{\frac{s^2 C_1 R_1 C_2 R_2 R_3}{R_4} - \frac{s C_1 R_1 R_2}{R_5} + \frac{C_1 R_1}{C_3 R_6}}{s^2 C_1 C_2 R_1 R_2 + s(2C_1 R_1 + C_2 R_2) + 1} I_{in}. \tag{15}$$

For the BS output, V_{in2} is connected to ground while V_{in1} and V_{in3} are connected to V_{HP} and V_{LP} outputs respectively giving:

$$I_{BS} = I_{o1} = I_{o2} = -\frac{\frac{s^2 C_1 R_1 C_2 R_2 R_3}{R_4} + \frac{C_1 R_1}{C_3 R_6}}{s^2 C_1 C_2 R_1 R_2 + s(2C_1 R_1 + C_2 R_2) + 1} I_{in}. \tag{16}$$

Therefore, the tAP and BS outputs are produced as current signals at the X_1 and X_2 terminals of $VCII_4$. In addition, as shown in Figure 4, both AP and BS outputs are available as voltage signals at the Z_1 and Z_2 terminals of $VCII_4$ in inverting and non-inverting forms, respectively.

From Equation (9) the quality factor Q and natural frequency ω_0 are determined using the following formulas:

$$\omega_0 = \sqrt{\frac{1}{C_1 C_2 R_1 R_2}}. \tag{17}$$

$$Q = \frac{\sqrt{C_1 C_2 R_1 R_2}}{2C_1 R_1 + C_2 R_2}. \tag{18}$$

By assuming $R_1 = R_2 = R$ and $C_1 = C_2 = C$ in Equations (17) and (18), we have:

$$\omega_0 = \frac{1}{RC}. \tag{19}$$

$$Q = \frac{1}{3}. \tag{20}$$

Therefore, ω_0 can be set by R and C independent of Q.

The sensitivities of the proposed filter are calculated using the well-known definition of sensitivity in Equation (15) [22] as:

$$S_x^F = \frac{x}{F}\frac{\partial F}{\partial x}. \tag{21}$$

$$S_{R_1}^{\omega_0} = S_{R_2}^{\omega_0} = S_{C_1}^{\omega_0} = S_{C_2}^{\omega_0} = -\frac{1}{2}. \tag{22}$$

$$S_{C_1}^{Q} = S_{R_1}^{Q} = \left[\frac{1}{2} - \frac{2R_1C_1}{(2C_1R_1 + C_2R_2)}\right]. \tag{23}$$

$$S_{C_2}^{Q} = S_{R_2}^{Q} = \left[\frac{1}{2} - \frac{R_2C_2}{(2C_1R_1 + C_2R_2)}\right]. \tag{24}$$

For $C_1 = C_2 = C$, $R_1 = R_2 = R$, from Equations (23) and (24), we have:

$$S_{C_1}^{Q} = S_{R_1}^{Q} = \left[\frac{1}{2} - \frac{2}{3}\right] = -\frac{1}{6}. \tag{25}$$

$$S_{C_2}^{Q} = S_{R_2}^{Q} = \left[\frac{1}{2} - \frac{1}{3}\right] = \frac{1}{6}. \tag{26}$$

As seen from Equations (22), (25), and (26), the proposed filter exhibits reduced sensitivities to passive parameters.

4. Non-Ideal Analysis

The non-ideal analysis of the proposed universal filter can be performed by considering the non-ideal current and voltage gains of the used VCIIs. Using Equation (1), we have:

$$I_2 = A_{IB}I_1; \; I_4 = A_{I1}I_3 \tag{27}$$

where A_{IB} and A_{I1} are the gain of the current buffer and the current gain of VCII$_1$ at its X$_1$ terminal, respectively.

At the input node (node 1) we have:

$$i_{in} = I_1 + A\prime_{I2}I_5, \tag{28}$$

where A'_{I2} is the current gain of VCII$_2$ at its X$_2$ terminal.

Using Equations (27) and (28) and Equations (2) and (4), I_2 is found as:

$$I_2 = \frac{A_{IB}(1 + sC_1R_1)(1 + sC_2R_2)}{s^2C_1C_2R_1R_2 + s(C_1R_1\{1 + A'_{I2}A_{I1}A_{IB}\} + C_2R_2) + 1}i_{in}. \tag{29}$$

From Equations (27) and (29), I_5 is obtained as:

$$I_5 = \frac{A_{IB}A_{I1}sC_1R_1}{s^2C_1C_2R_1R_2 + s(C_1R_1\{1 + A'_{I2}A_{I1}A_{IB}\} + C_2R_2) + 1}i_{in}. \tag{30}$$

V_{LP} can be expressed as:

$$V_{LP} = V_{Z23} = A''_{V2}V_{X23} = -A''_{I2}A''_{V2}\frac{1}{sC_3}I_5, \tag{31}$$

where V_{Z23} and V_{X23} are the voltages at the Z$_2$ and X$_2$ terminals of VCII$_3$, respectively. A''_{V2} is the voltage gain at the Z$_2$ terminal of VCII$_3$. Inserting Equation (30) into Equation (31), V_{LP} is found as:

$$V_{LP} = -\frac{A''_{I2}A''_{V2}A_{IB}A_{I1}\frac{C_1R_1}{C_3}}{s^2C_1C_2R_1R_2 + s(C_1R_1\{1 + A'_{I2}A_{I1}A_{IB}\} + C_2R_2) + 1}i_{in}. \tag{32}$$

From Equations (2) and (28), I_4 is found as:

$$I_4 = \frac{A_{IB}A_{I1}sC_1R_1(1+sC_2R_2)}{s^2C_1C_2R_1R_2 + s(C_1R_1\{1 + A'_{I2}A_{I1}A_{IB}\} + C_2R_2) + 1}i_{in}. \tag{33}$$

Inserting Equations (30) and (33) into Equation (11) gives I_{HP} as:

$$I_{HP} = \frac{-A_{IB}A_{I1}s^2C_1R_1C_2R_2}{s^2C_1C_2R_1R_2 + s(C_1R_1\{1 + A'_{I2}A_{I1}A_{IB}\} + C_2R_2) + 1}i_{in}. \tag{34}$$

From Equation (34), V_{HP1} and V_{HP2} are found as:

$$V_{HP1} = \frac{-A_{IB}A_{I1}A_{V2}s^2C_1R_1C_2R_2R_3}{s^2C_1C_2R_1R_2 + s(C_1R_1\{1 + A'_{I2}A_{I1}A_{IB}\} + C_2R_2) + 1}i_{in}. \tag{35}$$

$$V_{HP2} = \frac{-A_{IB}A_{I1}A''_{V2}s^2C_1R_1C_2R_2R_3}{s^2C_1C_2R_1R_2 + s(C_1R_1\{1 + A'_{I2}A_{I1}A_{IB}\} + C_2R_2) + 1}i_{in}. \tag{36}$$

Using Equations (33), (7), and (1), V_{BP} is found as:

$$V_{BP} = \frac{A_{IB}A_{I1}A_{V1}sC_1R_1}{s^2C_1C_2R_1R_2 + s(C_1R_1\{1 + A'_{I2}A_{I1}A_{IB}\} + C_2R_2) + 1}i_{in}, \tag{37}$$

where A_{v1} is the voltage gain at the X_1 terminal of VCII$_1$. From Equation (37), the quality factor Q and the natural frequency ω_0 are determined using the following formulas:

$$\omega_0 = \sqrt{\frac{1}{C_1C_2R_1R_2}}. \tag{38}$$

$$Q = \frac{\sqrt{C_1C_2R_1R_2}}{(C_1R_1\{1 + A'_{I2}A_{I1}A_{IB}\} + C_2R_2)}. \tag{39}$$

As seen from Equations (38) and (39), ω_0 is not affected by the non-ideal gains of VCIIs. In addition, as current and voltage gains of VCIIs are designed to be very close to unity, their effect on Q is also negligible. The relations for AP and BS outputs can be simply achieved by inserting Equation (32) and Equations (35)–(37) into Equation (14).

5. Simulation Results and Comparative Analysis

The proposed multifunction filter of Figure 3 has been simulated with PSpice and using 0.18 μm CMOS parameters under a supply voltage of ±0.9 V. The circuit of Figure 2 is used as a VCII. For I-CB, the current buffer section of Figure 2, consisting of transistors M_1–M_9 and current sources I_{B1}–I_{B3}, is used. The chosen transistor aspect ratios are shown in Table 1. The values of bias currents are $I_{B1} = 50$ μA, $I_{B2} = I_{B3} = I_{B4} = I_{B5} = I_{B6} = 20$ μA. The values of the passive components are: $R_1 = R_2 = 10$ KΩ, $C_1 = C_2 = 10$ pF. The values of R_3 and R_L are 5 KΩ.

Table 1. Transistor aspect ratios.

M_1–M_2	18/0.36	M_8–M_9	9/0.9
M_3–M_4	18/0.9	M_{13}	18/0.9
M_5	9/0.36	M_{10}, M_{15}	9/0.36
M_6–M_7	18/0.9	M_{11}, M_{14}	0.9/0.36

Figure 5a shows the frequency performances of the BP, HP, and LP outputs. The value of ω_0 and Q is measured as 1.55 MHz and 0.33, respectively. From Equations (17) and (18), the values of ω_0 and Q are calculated as 1.6 MHz and 0.33, respectively. As seen, there is a good agreement between theory and simulation results. The transient response of the different outputs is evaluated by applying a sinusoidal input current with a peak-to-peak value of 10 μA and frequency of 1.5 MHz. The total harmonic distortion (THD) values are 3.7%, 2.6%, and 3.6%, for HP, BP, and LP outputs, respectively.

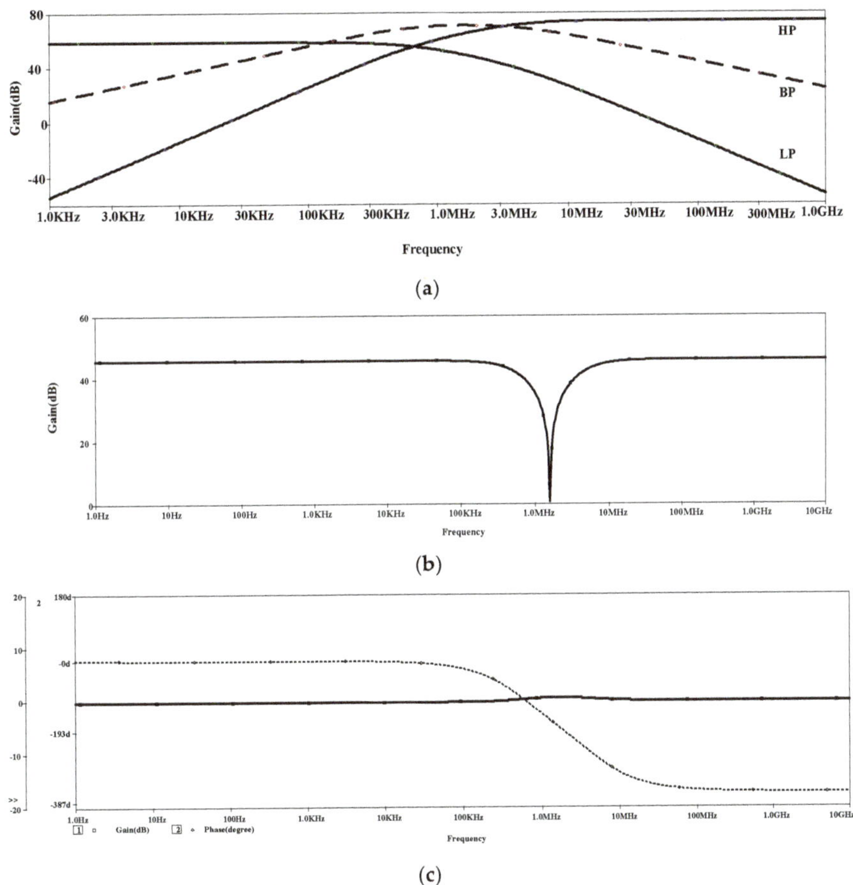

Figure 5. Frequency performance of the proposed universal filter for (**a**) band-pass (BP), high-pass (HP), and low-pass (LP) and (**b**) BS and (**c**) AP outputs.

Figure 5b shows the BS output, which is achieved by setting $R_4 = 12.5$ KΩ, $R_5 = 78$ kΩ, and $R_6 = 5$ KΩ and connecting $V_{in1} = V_{HP}$, $V_{in2} = 0$ and $V_{in3} = V_{LP}$. Figure 5c shows the gain and phase response of the AP output, which is achieved by connecting $V_{in1} = V_{HP}$, $V_{in2} = V_{BP}$ and $V_{in3} = V_{LP}$ and using the same values for R_4–R_6.

The tunability of the proposed filter is investigated by varying the center frequency ω_0, through C_1 and C_2. As shown in Figure 6, ω_0 varies from 1 to 32 MHz for different values of capacitors without affecting Q.

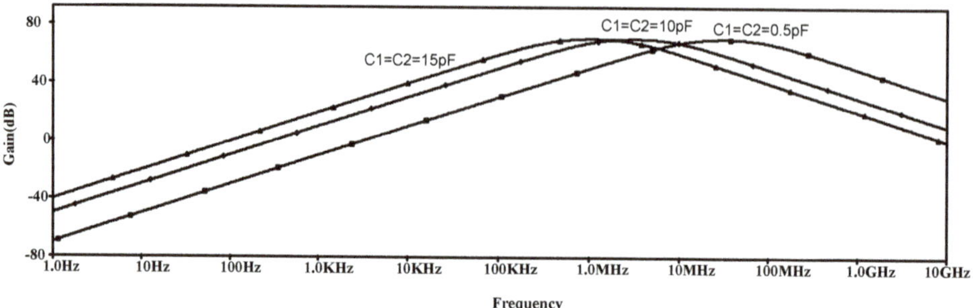

Figure 6. Natural frequency (ω_0) variation with C_1 and C_2.

The transient response of the different outputs is evaluated by applying a sinusoidal input current with a peak-to-peak value of 10 μA and a frequency of 1.5 MHz. The THD values are 2%, 1%, 6%, 1.4%, and 2% for the HP, BP, LP, AP, and BS outputs, respectively.

To investigate the circuit performance against transistor parameter tolerances, the circuit is simulated by applying 3% and 5% tolerance in V_{TH} and β (with usual meaning of symbols) of transistors. The result is shown in Table 2, which acknowledges the negligible effect of tolerances in the circuit performance. The input impedance of the proposed filter is only 47 Ω and the output impedances for voltage outputs and current outputs are 93 Ω and 254 KΩ, respectively. The power consumption is also 1.47 μW.

Table 2. Filter parameters vs. transistor parameters variation.

Variation in V_{TH} and β of Transistors	ω_0	Q
3%	1.68 MHz	0.32
5%	1.7 MHz	0.328

A comparison between the proposed universal filter and some other previously reported works is drawn in Table 3. As seen, the proposed circuit is the only one providing both inverting and non-inverting type AP and BS outputs in current and voltage forms. The proposed circuit is also the only one that can produce all possible outputs as current and voltage signals in low-impedance and high-impedance terminals, respectively. It also enjoys low-voltage operation. Although the circuit proposed in [7] does not employ floating capacitors, it requires three extra current buffers at the input. In addition, compared to the previously published universal filters of [6,11,14–17], which employ two floating capacitors, there is only one floating capacitor in the proposed circuit. Compared to [9,16,17], the proposed circuit does not require extra current and/or voltage buffers at the input and output terminals.

Table 3. Comparison between the proposed multifunction filter and other works.

		# of Floating Capacitors	Used Active Building Block	Vss-Vdd (V)	Pd (mW)	Outputs	Configuration	I–O Signals	ω_0	Extra Voltage Buffer
[6]	1	0	FTFN, OTA	NA	NA	LP, BP, BS	SIMO	V–V	7.95 KHz	Yes
	2	2	FTFN, OTA	NA	NA	LP, HP	SIMO	I–I	7.95 KHz	-
[7]		0	DVCC	±0.9	0.462	BP, BS, HP	MIMO	I–I	3.18 MHz	-*
[8]		0	CCCCTA	±1.85	NA	LP, BP, HP	SIMO	I–I	1.5 MHz	-
[9]	1	0	CFOA	NA	NA	LP, BP, HP	SIMO	I–I	<1 MHz	Current buffer required at output
	2	0	CFOA	NA	NA	LP, BP	SIMO	V–V	<1 MHz	Yes
[10]		0	CDBA	±1.25	NA	LP, BP, HP	SIMO	I–V	10 MHz	No
[11]		2	CDBA	±5	NA	AP, LP, BP	SIMO	I–V	1 MHz	No
[12]		0	MCCCDTA	±3	NA	LP, BP, HP, BS, AP	MISO	I–I	1.27 MHz	-
[14]		2	LT1228	±5	NA	LP, BP, HP, BS, AP	MISO	V–V	159.19 KHz	No
[15]		2	FDCCII	±1.65	NA	LP, BP, HP, BS, AP	MISO	V–V	1 MHz	No
[16]		2	CFCTA	±5	NA	LP, BP, HP, BS, AP	MISO	V–V	2.8 MHz	Yes
[17]		2	I-CB VDDDA	±1.25	NA	LP, BP, HP, BS, AP	MISO	V–V	1.074 MHz	Yes
This Work		1	I-CB, VCII$^\pm$	±0.9	1.47	HP, BP, LP I-AP, NI-AP, I-BS, NI-BP	SIMO	I/V–V/I	1–32 MHz	No

LP: Low pass; BP: Band pass; BS: Band stop; HP: High pass; AP: All pass, I: Inverting, NI: Non-inverting; SIMO: Single-input multiple-output; MIMO: Multiple-input multiple-output; MISO: Multiple-input single-output; FTFN: Four-terminal floating nullor; OTA: Operational transconductance amplifier; DVCC: Differential voltage current conveyor; CCCCTA: Current-controlled current conveyor transconductance amplifier; CFOA: Current feedback operational amplifier; CDBA: Current differencing buffered amplifier; MCCCDTA: Modified current-controlled current differencing transconductance amplifier; CFCTA: Current follower cascaded transconductance amplifier; I-CB: Inverting current buffer; VCII: Second-generation voltage conveyor. * The circuit suffers from high input impedances and three current buffers are required at inputs.

6. Conclusions

In this paper a new VCII-based multifunction filter was presented. It was demonstrated that, taking advantage from one current buffer and four second-generation voltage conveyors (VCII) as active building blocks, it can perform BP, HP, LP AP, and BS filtering actions simultaneously maintaining a very low circuit complexity as well as a low static power consumption. It was shown that, the versatility of the VCII block allows us to produce outputs in forms of both current and voltage signals. Filter functionality was also acknowledged through simulations, which showed a good agreement with the presented theory. The robustness of the circuit against fabrication mismatches was analyzed and a final comparison between the presented work and the available literature was given.

Author Contributions: L.S. designed the circuit and wrote the paper; G.B. performed the simulations; G.F. and V.S. have analyzed the equations and edited the paper.

Funding: This research received no external funding.

Conflicts of Interest: The authors declare no conflict of interest.

References

1. Alexander, C.; Sadiku, M. *Fundamental of Electric Circuits*, 2nd ed.; McGraw-Hill: New York, NY, USA, 2004.
2. Ibrahim, M.; Minaei, S.; Kuntman, H. A 22.5 MHz current-mode KHN-biquad using differential voltage current conveyor and grounded passive elements. *AEU Int. J. Electron. Commun.* **2005**, *59*, 311–318. [CrossRef]
3. Stornelli, V.; Pantoli, L.; Leuzzi, G.; Ferri, G. Fully differential DDA-based fifth and seventh order Bessel low pass filters and buffers for DCR radio systems. *Analog Integr. Circuits Signal Process.* **2013**, *75*, 305–310. [CrossRef]
4. Alzaher, H.; Al-Ghamdi, M.; Ismail, M. CMOS low-power bandpass IF filter for Bluetooth. *IET Circuits Devices Syst.* **2007**, *1*, 7–12. [CrossRef]
5. Prommee, P.; Saising, E. CMOS-based high-order LP and BP filters using biquad functions. *IET Circuits Devices Syst.* **2018**, *12*, 326–334. [CrossRef]
6. Shah, N.; Malik, M. High impedance voltage- and current-mode multifunction filters. *AEU Int. J. Electron. Commun.* **2005**, *59*, 262–266. [CrossRef]
7. Chen, H. Tunable versatile current-mode universal filter based on plus-type DVCCs. *AEU Int. J. Electron. Commun.* **2012**, *66*, 332–339. [CrossRef]
8. Singh, S.; Maheshwari, S.; Mohan, J.; Chauhan, D. Electronically Tunable Current-Mode Universal Biquad Filter Based on the CCCCTA. In Proceedings of the International Conference on Advances in Recent Technologies in Communication and Computing, Kottayam, India, 27–28 October 2009; pp. 424–429.
9. Horng, J.; Hou, C.; Huang, W.; Yang, D. Voltage/Current-Mode Multifunction Filters Using One Current Feedback Amplifier and Grounded Capacitors. *Circuits Syst.* **2011**, *2*, 60–64. [CrossRef]
10. Arora, T.; Rana, U. Multifunction Filter Employing Current Differencing Buffered Amplifier. *Circuits Syst.* **2016**, *7*, 543–550. [CrossRef]
11. Sagbas, M.; Koksal, M. A New Multi-mode Multifunction Filter Using CDBA. In Proceedings of the European Conference on Circuit Theory and Design, Cork, Ireland, 28 August–2 September 2005.
12. Kumngern, M. Current-Mode Multifunction Biquadratic Filter Using a Single MCCCDTA. In Proceedings of the IEEE International Conference on Cyber Technology in Automation, Control, and Intelligent Systems (CYBER), Bangkok, Thailand, 27–31 May 2012; pp. 115–118.
13. Kamat, D.; Ananda Mohan, P.; Gopalakrishna Prabhu, K. Current-mode operational transconductance amplifier-capacitor biquad filter structures based on Tarmy–Ghausi active-RC filter and second-order digital all-pass filters. *IET Circuits Devices Syst.* **2010**, *4*, 346–364. [CrossRef]
14. Klungtong, S.; Thanapatay, D.; Jaikla, W. Three-Input Single-Output Voltage-Mode Multifunction Filter with Electronic Controllability Based on Single Commercially Available IC. *Act. Passiv. Electron. Compon.* **2017**, *2017*, 5240751. [CrossRef]
15. Kumngern, M.; Supavarasuwat, P. A Voltage-Mode Three-Input One-Output Multifunction Filter Using Single FDCCII. In Proceedings of the 20th Asia-Pacific Conference on Communication (APCC2014), Pattaya, Thailand, 1–3 October 2014.
16. Chanapromma, C.; Jaikla, W.; Chaichana, A. New Realization of Single CFCTA-Based Voltage-Mode Multifunction Filter. In Proceedings of the 3rd International Conference on Control and Robotics Engineering, Nagoya, Japan, 20–23 April 2018.
17. Chaichana, A.; Sangyaem, S.; Jaikla, W. Multifunction Voltage-Mode Filter Using Single Voltage Differencing Differential Difference Amplifier. *MATEC Web Conf.* **2017**, *95*, 14003. [CrossRef]
18. Ferri, G.; Stornelli, V. A 0.18 μm CMOS DDCCII for Portable LV-LP Filters. *Radioengineering* **2013**, *22*, 434–439.
19. Safari, L.; Barile, G.; Ferri, G.; Stornelli, V. High performance voltage output filter realizations using second generation voltage conveyor. *Int. J. RF Microw. Comput.-Aided Eng.* **2018**, *28*, e21534. [CrossRef]
20. Safari, L.; Barile, G.; Stornelli, V.; Ferri, G. An Overview on the Second Generation Voltage Conveyor: Features, Design and Applications. *IEEE Trans. Circuits Syst. II Express Briefs* **2018**, *66*, 547–551. [CrossRef]

21. Carvajal, R.; Ramirez-Angulo, J.; Lopez-Martin, A.; Torralba, A.; Galan, J.; Carlosena, A.; Chavero, F. The flipped voltage follower: A useful cell for low-voltage low-power circuit design. *IEEE Trans. Circuits Syst. I Regul. Pap.* **2005**, *52*, 1276–1291. [CrossRef]
22. Chen, H. Single FDCCII-based universal voltage-mode filter. *AEU Int. J. Electron. Commun.* **2009**, *63*, 713–719. [CrossRef]

 *electronics*

Article

High-Performance Low-Pass Filter Using Stepped Impedance Resonator and Defected Ground Structure

Jin Zhang [1,2,3,*], Ruosong Yang [2] and Chen Zhang [2]

[1]   School of Electronic and Information Engineering, Jinling Institute of Technology, Nanjing 211169, China
[2]   National and Local Joint Engineering Laboratory of RF Integration and Micro-Assembly Technology, Nanjing 210023, China; 1018020806@njupt.edu.cn (R.Y.); 1018020802@njupt.edu.cn (C.Z.)
[3]   School of Electrical Engineering, Southeast University, Nanjing 210096, China
*   Correspondence: jinzhang@jit.edu.cn; Tel.: +86-298-618-8701

Received: 26 January 2019; Accepted: 1 April 2019; Published: 4 April 2019

Abstract: A microstrip low-pass filter (LPF) using reformative stepped impedance resonator (SIR) and defected ground structure (DGS) is proposed in this paper. The proposed filter not only possesses the advantage of high frequency selectivity of SIR hairpin LPF with internal coupling, but also possesses the large stop-band (SB) bandwidth by adjusting the number and area of DGS units. The LPF proposed in this paper possesses the properties of miniaturization, wide SB, high selectivity, and low pass-band ripple (PBR) simultaneously. The characteristic parameters of the proposed LPF is that: the pass-band (PB) is 0~2 GHz, the PBR is 0.5 dB, the SB range is from 2.4 GHz to 9 GHz when the attenuation is under 20 dB, and the maximal attenuation could reach 45 dB in the SB. The size of this proposed LPF is $0.13\lambda \times 0.09\lambda$; λ is the corresponding wavelength of the upper PB edge frequency of 2 GHz.

Keywords: low-pass filter (LPF); stepped impedance resonator (SIR); hairpin resonator; internal coupling; defected ground structure (DGS)

1. Introduction

In recent years, the miniaturized microstrip low-pass filter (LPF) with high frequency selectivity and wide stop-band (SB) is widely used in a RF/microwave circuit such as satellite and mobile communication systems. However, based on step impedance resonator (SIR), the traditional microstrip LPFs implemented by Richard transform and Kuroda rule can only constitute the LPFs of the type of Butterworth and Chebyshev responses [1]. The responses of these two filters with single-sections exhibit poor frequency selectivity. Therefore, in order to obtain a steep selecting edge, the number of filter sections must be increased [2,3]. At the same time, the insertion loss in the pass-band (PB) and the physical size of the filter will be increased. On the basis of a previous study [4], Reference [5] proposed a PCB LPF using a semi-lumped parallel resonance circuit which exhibits an elliptic function response. In Reference [6], the coupling between two LP hairpin filters was performed to obtain a bandpass response. These kinds of filters achieve high frequency selectivity, but the SB width is not wide enough, and the attenuation out of the band is insufficient.

In order to realize a microstrip LPF with the characteristics of miniaturization, high selectivity and wide SB suppression at the same time, the research hotspot of the defected ground structure (DGS) of nearly two decades is combined with the SIR hairpin filter in this paper. The DGS was proposed to design LPF by Korean scholars D. Ahn et al. in 2001 [7]. From then on, more and more research findings have used DGS to design a high performance microstrip filter. Nechel et al. [8] introduced metamodels in filter design by use of a DGS. The optimal time of the design process decreases drastically, however, the simple slot DGS analyzed in the reference is not suitable for filter design for high frequency selectivity. In Reference [9], asymmetrical Pi-shaped DGSs with Koch fractal curve were used to design compact LPF, however, the multiple asymmetrical fractal structures undoubtedly increase the filter's size. Zeng et al. [10] used compact DGSs with mutual capacity and inductive coupling to broaden the bandwidth and the rejection ratio in SB range; the shortcoming of this tight coupling design is that it increases the PB insertion loss. A compact LPF was designed by combining several fractal DGSs with a conventional multiple stub filter in Reference [11]. However, the size of the filter designed in the reference appears insufficiently compact due to the usage of conventional multiple stub LPF in the top layer.

In this paper, we use the SIR hairpin resonator with internal coupling to make the filter more compact and use the properties of the DGS to widen the SB by optimization design. This paper is organized as follows. In Section 2, the electrical characteristics of one-, two- and three-DGS are investigated. A compact LPF with high frequency selectivity is proposed based on SIR hairpin with internal coupling in Section 3. A miniaturization, high selectivity and wide SB suppression microstrip LPF are simulated and experimentally validated in Section 4. Finally, Section 5 gives some concluding remarks.

2. The Electrical Properties of DGS

The DGS etched on the grounding metal plate for the microstrip circuit can be square-headed, semi-circular-headed dumbbell-shaped, and snake-shaped patterns. In this paper, the square-headed dumbbell-shaped DGS is applied. The 3-D model of one-, two- and three-DGS structures are shown in Figure 1a–c.

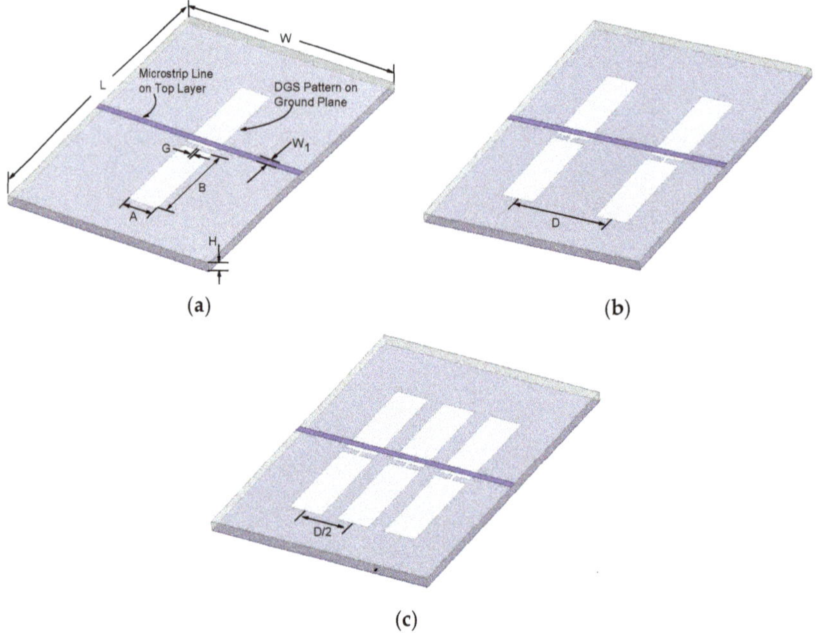

Figure 1. 3-D model of (**a**) one-, (**b**) two- and (**c**) three-DGS structures.

The model of one-DGS structure in Figure 1a is simulated in Ansoft HFSS. This model is designed on PTPE composite medium copper-clad foil plate with a relative dielectric constant $\varepsilon_r = 10.8$ and a loss tangent of 0.0035. The distance of the middle slots is represented by G, the linewidth of microstrip line by W_1 for 50-Ω characteristic impedance, and the width of DGS by A. The size of the whole model is about $L \times W \times H$. The detailed numerical values of the model are listed in Table 1. Figure 2a shows the simulated frequency responses of the return loss S_{11} and the insertion loss S_{21} of one-DGS case when the length B of DGS are 5.8, 6.8, 7.8, and 8.3 mm, respectively.

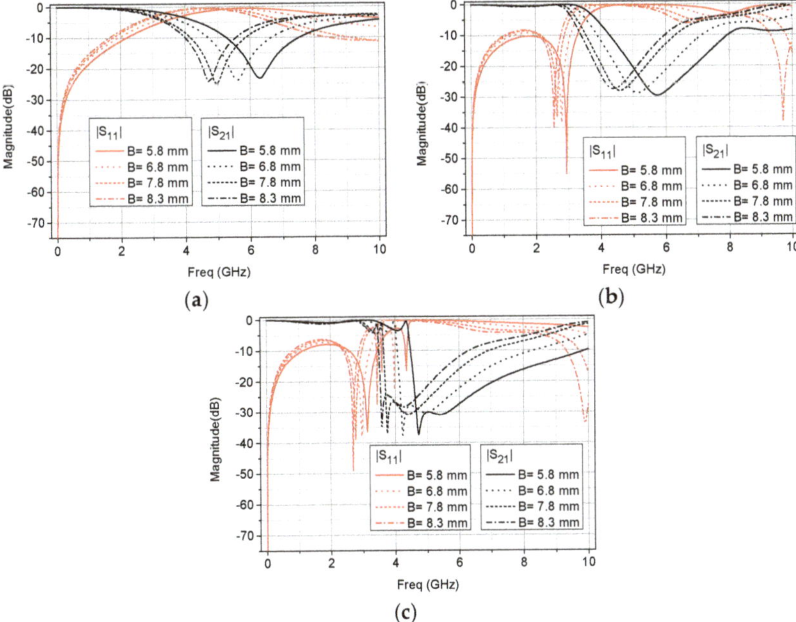

Figure 2. Frequency responses of return loss and insertion loss of the model of (**a**) one-DGS, (**b**) two-DGS and (**c**) three-DGS.

Table 1. Filter specifications.

Dielectric Substrate	Values	Model	Values (mm)
ε_r	10.8	L	20
Loss tangent	0.0035	W	13.8
		H	0.5

SIR hairpin filter	Values (mm)	DGS	Values (mm)
W_1, W_2, W_3	0.396, 0.232, 0.9	G	0.2
L_1	5.92	A	2
L_2, L_3, L_4, L_5	1, 2.626, 0.796, 6.3	D	6
S	0.19	B_1, B_2	7.8, 8.3

The transmission zero of the one-DGS structure shifts to the lower frequency band with the increasing length of B as shown in Figure 2a. The frequencies of the transmission zero are 6.28 GHz, 5.62 GHz, 4.96 GHz, and 4.69 GHz, respectively. Further study shows that the frequencies of transmission zeros decrease with the increasing areas of DGS. Actually, the electrical property of the model of one DGS under the microstrip is equivalent to that of a parallel L-C resonator circuit. The growing area of the DGS corresponds to increasing the equivalent inductance of the parallel resonance circuit; the transmission zero therefore moves to the low frequency with increasing area of the DGS.

Similarly, the simulation frequency response characters of the models consisted of two- and three-DGS units which are presented in Figure 1b,c and Figure 2b,c, respectively. The distance between adjacent DGSs for two- and three-DGS models are D/2 = 3 mm. The sizes of the other parts are set up identical to the one-DGS case.

The transmission zeros of two- and three-DGS structures also move to the low frequency band with the increasing length of B as shown in Figure 2b,c. By comparing the Figure 2a–c, the frequency positions of transmission zeros decrease with the increasing number of DGS, and the width of the transition band (TB) between PB and SB greatly reduced in the three-DGS case. The interpretation of this phenomenon can be illustrated in Figure 3.

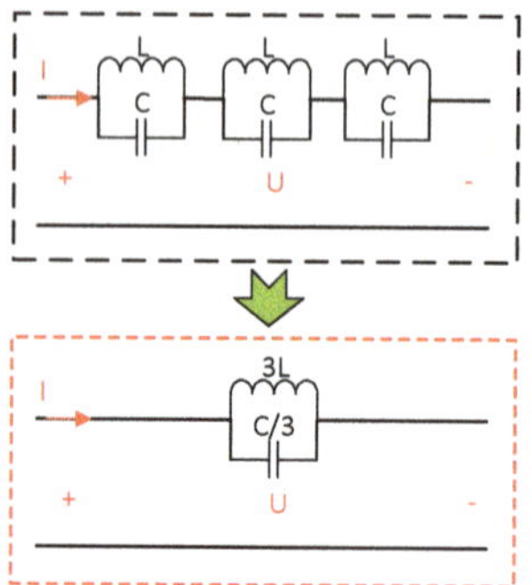

Figure 3. The equivalent circuits of the three-DGS model in Figure 2c.

The equivalent circuit model of the three-DGS model can be modeled by three parallel L-C resonators, as shown in the black dashed box in Figure 3. The mutual coupling effects between each resonator are ignored because of a greater distance between the adjacent DGSs. The relations of the electric parameters for this circuit can be expressed as

$$U = I\left(3 \cdot \frac{1}{\frac{1}{j\omega L} + j\omega C}\right) \tag{1}$$

Via simple mathematical transformation, Equation (1) converts into Equation (2)

$$U = I\left[\frac{1}{\frac{1}{j\omega(3L)} + j\omega(C/3)}\right] \tag{2}$$

The single equivalent parallel L-C resonator in the red dotted line in Figure 3 reflects the relations of the electric parameters in Equation (2). Because the area of the DGS mainly affects the equivalent inductance of the parallel resonator, the three lengthening side lengths of B triple the parallel inductance. From DGS simulation in Figure 2, it is known that the structure of DGS has the characteristics of low-pass filtering. Its low-pass characteristics can be adjusted by changing the defected area and shape etched on the grounding plate. This characteristic is applied to broaden the SB width of the SIR hairpin LPF in Section 3.

3. Investigation of Proposed SIR Hairpin LPF

To realize the function of filtering, a filter requires one or more units of resonant structure. The resonant elements of SIR structure for filter designing is an effective way to realize compact filters. Lung-Hwa Hsieh et al. proposed a LPF using a SIR hairpin unit [2]. On this basis, combined with M. Makimoto's monograph [12], the following LPF with high attenuation edges are proposed. The cell structure is shown in Figure 4.

The equivalent circuit of the resonator shown in Figure 4 can be regarded as a parallel connection of a single transmission line and two parallel lines with internal coupling. The characteristic impedance and electric length of the single transmission line are Z_S and θ_T, respectively. The characteristic impedance (the electric length) of the even and odd modes of the parallel lines with internal coupling are Z_{pe}, Z_{po} (θ_{pe}, θ_{po}), respectively. The A matrix of the single transmission line and the parallel coupling line are expressed as A_I and A_{II}, respectively.

$$A_I = \begin{bmatrix} A_1 & B_1 \\ C_1 & D_1 \end{bmatrix}$$

$$= \begin{bmatrix} \cos\theta_T & jZ_S\sin\theta_T \\ \dfrac{j\sin\theta_T}{Z_S} & \cos\theta_T \end{bmatrix} \tag{3}$$

$$A_{II} = \begin{bmatrix} A_2 & B_2 \\ C_2 & D_2 \end{bmatrix}$$

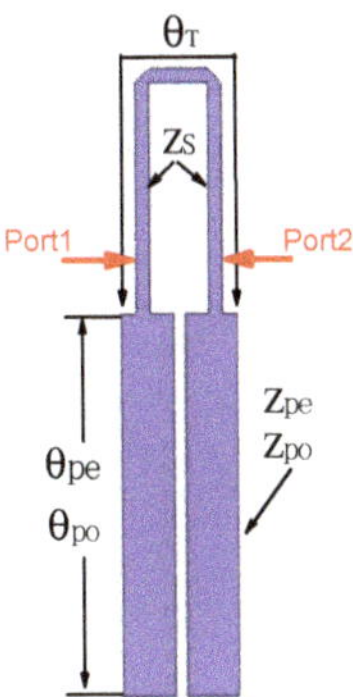

Figure 4. SIR hairpin with internal coupling.

$$= \begin{bmatrix} \dfrac{Z_{pe}\cot\theta_{pe}+Z_{po}\cot\theta_{po}}{Z_{pe}\cot\theta_{pe}-Z_{po}\cot\theta_{po}} & -j\dfrac{2Z_{pe}Z_{po}\cot\theta_{pe}\cot\theta_{po}}{Z_{pe}\cot\theta_{pe}-Z_{po}\cot\theta_{po}} \\ \dfrac{j2}{Z_{pe}\cot\theta_{pe}-Z_{po}\cot\theta_{po}} & \dfrac{Z_{pe}\cot\theta_{pe}+Z_{po}\cot\theta_{po}}{Z_{pe}\cot\theta_{pe}-Z_{po}\cot\theta_{po}} \end{bmatrix} \tag{4}$$

The matrix A_T is defined as the total A matrix of the above two parallel circuits.

$$A_T = \begin{bmatrix} A_t & B_t \\ C_t & D_t \end{bmatrix} \tag{5}$$

where

$$A_t = \frac{A_1 B_2 + A_2 B_1}{B_1 + B_2} \tag{6a}$$

$$B_t = \frac{B_1 B_2}{B_1 + B_2} \tag{6b}$$

$$C_t = \frac{-(A_2 - A_1)(D_2 - D_1) - (B_2 + B_1)(C_2 + C_1)}{B_1 + B_2} \tag{6c}$$

$$D_t = \frac{D_1 B_2 + D_2 B_1}{B_1 + B_2} \tag{6d}$$

where $D_t = A_t$ due to $D_1 = A_1$ and $D_2 = A_2$. A load impedance Z_L is connected to the terminal in the circuit. The input admittance Y_I can be calculated based on Equation (5):

$$Y_I = \frac{C_t + \frac{D_t}{Z_L}}{A_t + \frac{B_t}{Z_L}} \tag{7}$$

The resonance condition is met under $Y_I = 0$ and $Z_L = \infty$, therefore, substituting the calculated value in Equation (5), the resonance condition is obtained as follows:

$$\begin{aligned}(Z_{pe} \cdot Z_{po} \cot\theta_{pe} \cot\theta_{po} - Z_S^2)\sin\theta_T + Z_S(Z_{pe}\cot\theta_{pe} + Z_{po}\cot\theta_{po})\cos\theta_T \\ -Z_S(Z_{pe}\cot\theta_{pe} - Z_{po}\cot\theta_{po}) = 0\end{aligned} \tag{8}$$

Based on the resonance condition of Equation (8), the characteristics of the above resonator can be analyzed using the microwave simulator method [12]. Figure 5 shows the geometry of the resonator connected with two tapped microstrip lines and its equivalent circuit diagram.

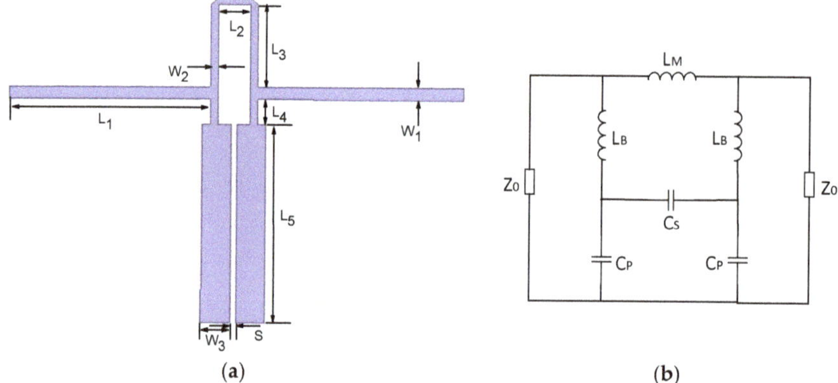

Figure 5. (a) Layout of LPF constructed by SIR hairpin with internal coupling; (b) equivalent circuit diagram of microstrip LPF.

The nonideal inductance L_M in Figure 5b is the equivalent inductance of L_3, L_2, L_3 and two 45° bends in Figure 5a. L_B is the equivalent inductance of L_4. The non-ideal capacitances C_S, C_P are equivalent to the capacitances generated by two open coupling lines, in which C_S is generated by the coupling effect of two lines; C_P is generated by the coupling between the terminal microstrip line and the ground plate; and Z_0 is the characteristic impedance of two feeding lines. The equivalent circuit in Figure 5b shows that the proposed structure has the property of an elliptic function-like LPF. The corresponding circuit model in Figure 5a is simulated by ADS software. The simulation is to design a LPF with a cut-off frequency of 2 GHz under the condition that the relative dielectric constant of the microstrip substrate is 10.8 and the thickness of the substrate is 0.5 mm. The simulated schematic in ADS software is shown in Figure 6a. To improve the frequency selectivity, the structural dimensions of the hairpin in Figure 5a are obtained through software optimizing measures, which are listed in Table 1. The transmission characteristics of the filters tuned by software are shown in Figure 6b.

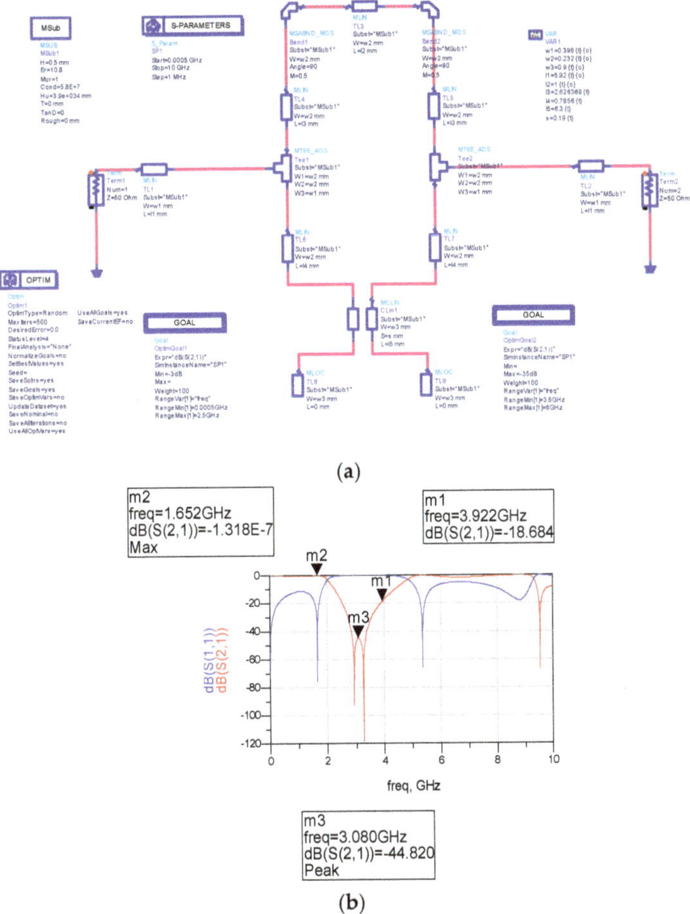

(a)

(b)

Figure 6. SIR hairpin LPF composed of single resonator with internal coupling. (**a**) The schematic diagram modeled in Agilent ADS software. (**b**) The simulation transmission characteristics.

As shown in Figure 6b, the proposed single SIR filter possesses the characteristic of high selectivity and produces two attenuation poles within a finite frequency band. This is the transmission characteristic of the three-branch elliptic function LPF. The disadvantage of this filter is that the SB width is narrow. Although the SB width can be properly enlarged by changing the size of each part of the SIR, it sacrifices the original characteristics of high frequency selectivity. Next, the SB width will be extended without changing the above structure sizes. That is to say, without changing the high edge frequency selectivity of the filter, the SB bandwidth is extended. The method is to introduce DGS structure.

4. Model Simulation and Sample Measurement

The proposed filter model is shown in Figure 7 using the design method of combining DGS with an SIR hairpin filter. The optimized dimensions of the SIR hairpin filter on the top layer are identical to those in Figure 5a. The sizes of the dielectric substrate and the three DGS patterns are the same with those in Figure 1c except that the length of the middle GDS B_2 is slightly longer than those of the two side GDS B_1. The optimized lengths of B_1 and B_2 are different in order to widen the valid SB width. The electrical parameters of the dielectric substrate and the size of each part of the filter are listed in Table 1.

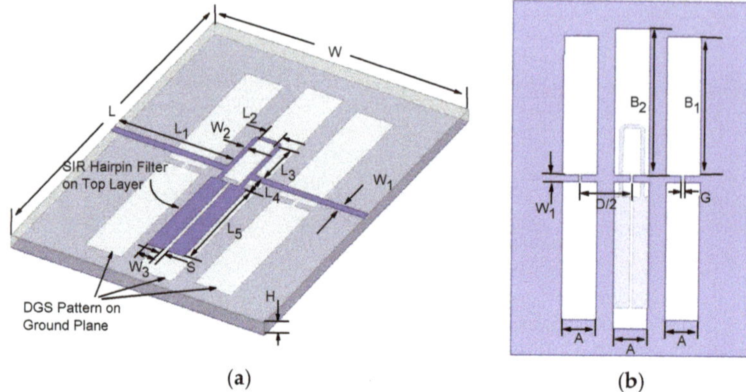

Figure 7. LPF model based on SIR and DGS. (**a**) 3-D model diagram and (**b**) DGS pattern diagram on the backside of the proposed filter.

The proposed filter is fabricated and shown in Figure 8. The Keysight N5224A PNA network analyzer is used to measure the S-parameters of the proposed LPFs. Figure 9 presents the simulated and measured results. The red lines are the S-parameters for a single SIR hairpin filter. These results obtained by full-wave electromagnetic simulation coincide with the results achieved by the circuit simulation using ADS software in Figure 6.

From the simulation results in Figure 9, it can be seen that the transmission coefficient of $S_{21,dB}$ is less than 0.38 dB from 0 Hz to 2 GHz, and the SB width is from 2.42 GHz to 8.2 GHz (see the blue curve) with a rejection of greater than 20 dB. Fortunately, the testing results show a broader SB width from 2.42 GHz to 9 GHz (see the black curve) with the same frequency selectivity and the attenuation in SB. The upper SB value even exceeds 10 GHz if the maximum attenuation degree is determined as 14.45 dB. The simulation and testing results are in good agreement.

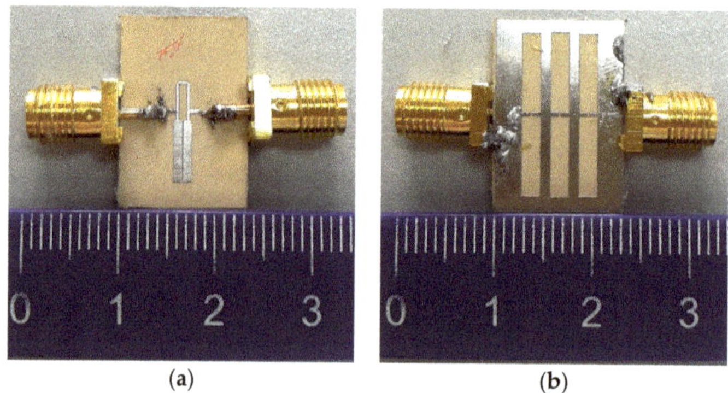

Figure 8. The fabricated microstrip LPF based on SIR and DGS: (**a**) top layer and (**b**) bottom layer.

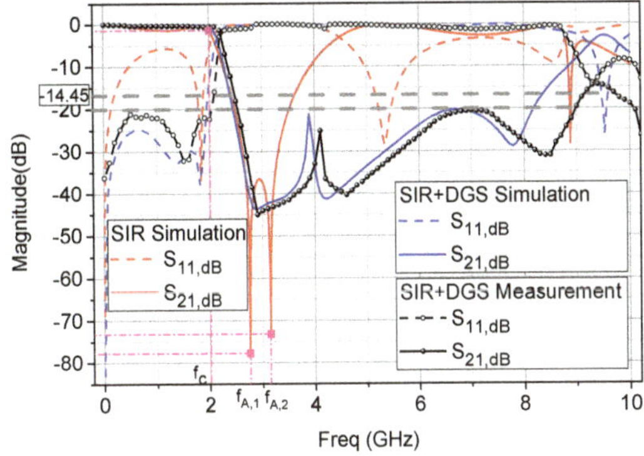

Figure 9. The simulation and measured results of S-parameters.

For the hairpin resonant filter, the simulated S-parameters shown in Figure 9 by the red curves can be used to obtain the input impedance Z_{IN} [1]:

$$Z_{IN} = \frac{1 + \left(S_{11} + \frac{S_{12}S_{21}\Gamma_L}{1 - S_{22}\Gamma_L}\right)}{1 - \left(S_{11} + \frac{S_{12}S_{21}\Gamma_L}{1 - S_{22}\Gamma_L}\right)} Z_0 \tag{9}$$

where $\Gamma_L = (Z_L - Z_0)/(Z_L + Z_0)$ is the load reflection coefficient, and Z_0 and Z_L are the characteristic impedance of the transfer line and the load impedance, respectively. The Z_0 and Z_L in our manuscript are also set to 50 Ω. Therefore, the $Z_{IN} = 50(1 + S_{11})/(1 - S_{11})$ and its values are shown in Figure 10. From the Z_{IN} in Figure 10b, it can be found that the first resonant frequency $f_{R,1} = 1$ GHz. Therefore, the SIR hairpin filter is a resonant element.

The quality-factor (Q) of a filter is the parameter to present the ratio of reactive power to active power. In the bandpass filter, the higher Q values mean the narrower bandwidth of the filter [13]. Based on Z_{IN}, the quality-factor (Q) value is given by $Q = \text{Im}(Z_{IN})/\text{Re}(Z_{IN})$. The Q values reach 1120, 1214 and 0.86 at the two attenuation poles $f_{A,1} = 2.75$ GHz, $f_{A,2} = 3.14$ GHz and the cut-off frequency $f_C = 2$ GHz, respectively. The Q values theoretically would be 0 at the resonant frequency points. The Q values at resonant frequency points $f_{R,1}, \cdots, f_{R,6}$ are very small values rather than zero, e.g., $Q_{f_{R,1}} = 0.0078$ at $f_{R,1}$. This is because the very accurate frequency points for resonance failed to take using the discrete digital computing method.

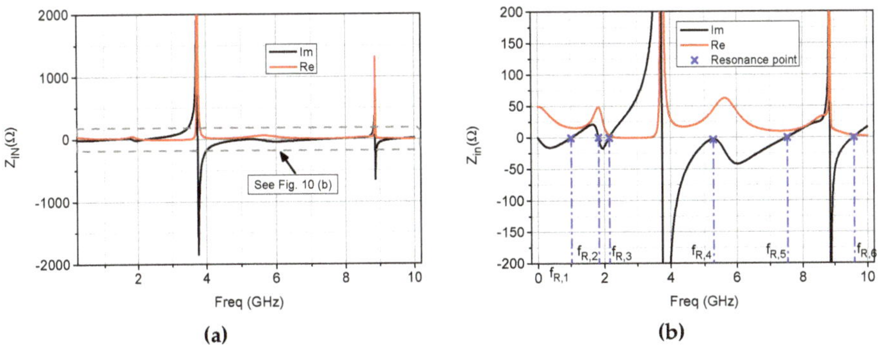

Figure 10. The input impedance of the SIR hairpin resonant filter. (**a**) Global view and (**b**) Partial view.

According to the S-parameters calculated in Figure 9, we present the insertion loss (IL) in decibel form based on the equation of $IL_{dB} = 10\lg|S_{21}|^2 = 2S_{21,dB}$ [1,14,15]. Therefore, the values of IL_{dB} are twice the values of $S_{21,dB}$ over the range of the investigated frequency.

The group delay as the derivative of the phase varies as an important parameter in microwave filter design. The smaller the group delay variation, the better the flat property of the designed filter. The simulation and measured values of group delay are shown in Figure 11. From this figure, the group delay varies from 0.23 to 0.78 ns in the pass band of 0~2 GHz. That is to say, the maximum variation of group delay is about 0.55 ns, representing a good flat property of the designed filter.

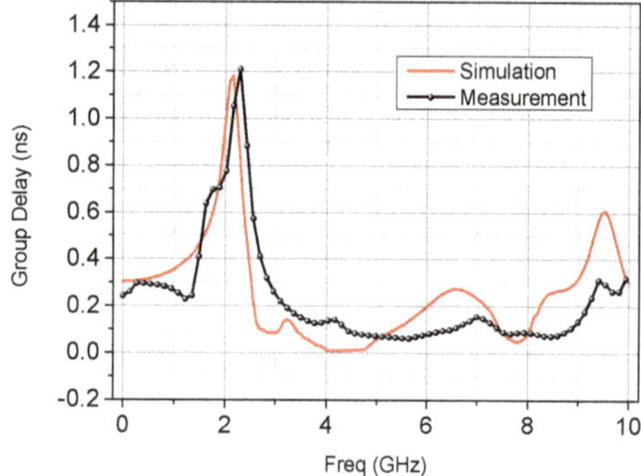

Figure 11. The simulation and measured values of group delay of the designed filter.

Some LPFs proposed in the references are compared with this paper in Table 2. The proposed LPF in this paper has smaller pass-band ripple (PBR) by comparing with the LPF in Reference [10]. The size of our designed LPF is less than those in Reference [9,11] with similar SB and TB width. The attenuation in stop-band (ASB) of LPF in Reference [2] is better than that of our design, however, the size of the designed filter in the reference is larger than our proposed filter in this paper. Therefore, the LPF proposed in this paper possesses the properties of miniaturization, wide SB, high selectivity, and low PBR simultaneously.

Table 2. Performance comparisons between the reported filters and the proposed filter in this paper (SB: stop-band, TB: transition band, PBR: pass-band ripple, ASB: attenuation in stop-band).

Ref.	Size/λ	SB (GHz)	TB (GHz)	PBR (dB)	Substrate
This work	0.13 × 0.09	6.6 *; >10 **	0.4 *; 0.35 **	≤0.5	PTPE composite medium
[2]	0.25 × 0.08	7.6 ***	0.4 ***	low	RT/Duroid 6010
[11]	0.42 × unavailable	6.5 *	0.5 *	low	ARLON 25N
[9]	0.13 × 0.23	6.7 *	0.3 *	low	FR4
[10]	0.1 × 0.1	7.4 *	0.2 *	high	FR4

* ASB < 20 dB; ** ASB < 14 dB; *** ASB < 30 dB.

5. Conclusions

Firstly, the square-headed dumbbell-shaped DGS is introduced. The transmission characteristics of one-, two- and three-DGS units are analyzed and compared by the 3D numerical simulation software HFSS. The results show that the structure of DGS has the property of low-pass filtering, and the SB attenuation extremum frequency decreases with the increase of the area and number of DGS units.

Secondly, a compact LPF is designed by using SIR hairpin resonator with internal coupling, which is equivalent to three-branch elliptic function LPF. The optimal sizes of the SIR LPF are obtained by the engineering electromagnetic software ADS of Agilent company. From the result of the simulation, this compact hairpin LPF has high frequency selectivity and very low PB attenuation, but the SB width under 20 dB is not wide enough.

Finally, to extend the SB rejection width, we combine the DGS with the SIR LPF. By adjusting the area and number of DGS units, a microstrip LPF with small size, high frequency selectivity and wide SB suppression is achieved. Meanwhile, the maximum group delay variation in the passband is within 0.55 ns. The manufactured compact filter keeps good consistency with the simulation results, which proves the validity of the theoretical design. This compact LPF with excellent transmission performance is very suitable for the application of modern wireless communication circuits.

Author Contributions: Conceptualization, J.Z.; Methodology, J.Z.; Software, J.Z.; Validation, J.Z.; Formal analysis, J.Z.; Investigation, C.Z.; Resources, J.Z.; Data curation, J.Z.; Writing—original draft preparation, R.-S.Y. and J.Z.; Writing—review and editing, J.Z.

Funding: This research was supported in part by the open research fund of the National and Local Joint Engineering Laboratory of RF Integration and Micro-Assembly Technology (KFJJ20180203), the Scientific Research Foundation for the High-Level Talents of Jinling Institute of Technology (jit-b-201719), and the Scientific Research Incubation Foundation of Jinling Institute of Technology (jit-fhxm-201802).

Acknowledgments: The authors would like to thank D. Chen for experimental support to this work.

Conflicts of Interest: The authors declare no conflict of interest.

References

1. Pozar, D.M. *Microwave Engineering*, 2nd ed.; Wiley: New York, NY, USA, 1998; Chapter 8.
2. Lung-Hwa, H.; Kai, C. Compact elliptic-function low-pass filters using microstrip stepped-impedance hairpin resonators. *IEEE Trans. Microw. Theory Tech.* **2003**, *51*, 193–199. [CrossRef]
3. Liu, X.; Wu, W.; Ji, P.; Yuan, N. Design of Compact Dual-Passband Filters by Parasitic Passband With Controllable Passbands. *IEEE Microw. Wirel. Compon. Lett.* **2018**, *28*, 410–412. [CrossRef]
4. Saal, R.; Ulbrich, E. On the Design of Filters by Synthesis. *IRE Trans. Circuit Theory* **1958**, *5*, 284–327. [CrossRef]
5. Sheen, J. A compact semi-lumped low-pass filter for harmonics and spurious suppression. *IEEE Microw. Guided Wave Lett.* **2000**, *10*, 92–93. [CrossRef]
6. Mora, S.; Alonso, Y.; Vargas, N.; Vera, J.; Avendano, J. Design of a bandpass filter using microstrip Hairpin resonators. In Proceedings of the 2017 CHILEAN Conference on Electrical, Electronics Engineering, Information and Communication Technologies (CHILECON), Pucon, Chile, 18–20 October 2017; pp. 1–5.
7. Ahn, D.; Park, J.; Kim, C.; Kim, J.; Qian, Y.; Itoh, T. A design of the low-pass filter using the novel microstrip defected ground structure. *IEEE Trans. Microw. Theory Tech.* **2001**, *49*, 86–93. [CrossRef]
8. Nechel, E.V.; Ferranti, F.; Latairc, J.; Rolain, Y. Efficient Design Optimization and Variability Analysis of Defected Ground Structure Filters Using Metamodels. In Proceedings of the 2017 IEEE MTT-S International Microwave and RF Conference (IMaRC), Ahmedabad, India, 11–13 December 2017; pp. 1–4.
9. Cao, S.; Han, Y.; Chen, H.; Li, J. An Ultra-Wide Stop-Band LPF Using Asymmetric Pi-Shaped Koch Fractal DGS. *IEEE Access* **2017**, *5*, 27126–27131. [CrossRef]
10. Zeng, Z.; Yao, Y.; Zhuang, Y. A Wideband Common-Mode Suppression Filter with Compact-Defected Ground Structure Pattern. *IEEE Trans. Electromagn. Compat.* **2015**, *57*, 1277–1280. [CrossRef]
11. Kufa, M.; Raida, Z. Lowpass filter with reduced fractal defected ground structure. *Electron. Lett.* **2013**, *49*, 199–201. [CrossRef]
12. Makimoto, M.; Yamashita, S. *Microwave Resonators and Filters for Wireless Communication Theory, Design and Application*; Springer: Berlin, Germany, 2001; Chapter 4.
13. Ahmadivand, A.; Sinha, R.; Gerislioglu, B.; Karabiyik, M.; Pala, N.; Shur, M. Transition from capacitive coupling to direct charge transfer in asymmetric terahertz plasmonic assemblies. *Opt. Lett.* **2016**, *41*, 5333–5336. [CrossRef] [PubMed]

14. Gerislioglu, B.; Ahmadivand, A.; Pala, N. Tunable plasmonic toroidal terahertz metamodulator. *Phys. Rev. B* **2018**, *97*, 161405. [CrossRef]

15. Kruger, B.A.; Joushaghani, A.; Poon, J.K.S. Design of electrically driven hybrid vanadium dioxide (VO$_2$) plasmonic switches. *Opt. Express* **2012**, *20*, 23598. [CrossRef] [PubMed]

 *electronics*

Article

Disturbance and Signal Filter for Power Line Communication

Krzysztof Bernacki [1,*], Dominik Wybrańczyk [1], Marcin Zygmanowski [2], Andrzej Latko [2], Jarosław Michalak [2] and Zbigniew Rymarski [1]

[1] Institute of Electronics, Faculty of Automatic Control, Electronics and Computer Science, Silesian University of Technology, 44-100 Gliwice, Poland; dominik.wybranczyk@gmail.com (D.W.); zbigniew.rymarski@polsl.pl (Z.R.)

[2] Department of Power Electronics, Electrical Drives and Robotics, Faculty of Electrical Engineering, Silesian University of Technology, 44-100 Gliwice, Poland; marcin.zygmanowski@polsl.pl (M.Z.); andrzej.latko@polsl.pl (A.L.); jaroslaw.michalak@polsl.pl (J.M.)

* Correspondence: krzysztof.bernacki@polsl.pl

Received: 23 December 2018; Accepted: 23 March 2019; Published: 28 March 2019

Abstract: Today, to use home automation, intelligent home controls or remote controls in the office, electronic equipment is moving away from wireless communication in favor of Power Line Communication (PLC). In the standard PLC solutions, the corrections that result from error transmissions are based on complex digital modulation methods and algorithms for validating the transmitted data without paying attention to the causes of the errors. This article focuses on the implementation of a filtering system for interference and signals in the 120–150 kHz band (CENELEC band C), which is injected into the network by transmitters. Such a filter separates the desired signal from the interference that is occurring in the network, which can result in communication errors. Moreover, when used properly, the filter can be used as a subsystem separation element. The paper presents the requirements, design, construction, simulation and test results that were obtained under actual operating conditions. It is possible to use less complex methods for correcting errors in transmission signals and to guarantee an improvement in the transmission rate using the proposed filter system.

Keywords: power line communication (PLC); conducted disturbances; anti-interference filter; smart home

---

1. Introduction

Power line communication systems have been used for various purposes within a broad range of fields, including telephony [1], street lighting controls, intercoms [2] and power transmission systems for almost 100 years [3,4]. Currently, Power Line Communications (PLCs) have a wide range of applications in smart grids [5,6]. The smart building market is in an early stage of its development and has great potential for future growth. According to the report "Smart Building Market Industry Analysis, Size, Growth, Trends, Segment and Forecast 2016–2021" that was prepared by MarketsandMarkets, the value of this market will increase from 7.26 billion dollars to 36.40 billion dollars by 2021 and the expected annual growth rate will be 38% [7]. The idea of an intelligent building has already taken root in developed Asian countries and in the USA, and its rapid development in Europe is also predicted. The fact that this trend is supported by a European Union policy is not without significance and the promotion of energy efficiency and pro-ecological solutions, including intelligent constructions, has been mentioned among the priorities for the coming years [8–11]. As the current market for intelligent home management systems is heavily dominated by companies that only sell their products to individuals with a high property status, smart home furnishings are an expression of luxury and prestige [12–14]. However, there are innovative systems that are becoming available at an affordable price, whose target group will be households with a good or medium financial standing, including both the owners of newly built houses/flats as well as existing homes/apartments.

Intelligent building systems come in three forms. The first is comprised of wired systems that use a dedicated communication cable whose big advantage is its stability. However, due to the use of this dedicated communication cable, these systems require a rethinking of the deployment and a definition of the desired functionalities at the beginning of the investment. Therefore, choosing this system for apartments in the secondary market requires a costly renovation. Other disadvantages are their complicated configuration and high price. An alternative solution is a wireless system [15]. The advantages of wireless systems include the ability to adapt the intelligent system to any building and to ensure price competitiveness. On the other hand, due to electromagnetic interference in the buildings as well as difficulties in data transmission in the event of obstacles in the signal path and directives that limit the maximum allowable transmitting powers, these wireless systems do not always work stably. Systems that work in a mesh topology can compensate for the effects of unstable operations; however, due to their characteristics, if one of the elements fails, the system may cease to function completely. It should also be noted that the use of wireless solutions may become problematic in the near future due to the number of devices that operate in a wireless mode. IoT solutions should also be distinguished [16]. However, IoT solutions may encounter the same problems that were mentioned earlier for wireless products. The third possible solution is wired systems that have no additional communication cable. Such PLC systems communicate via power lines; thus, they combine stability with ease of use and can easily be adapted to existing buildings. A hybrid solution that uses both wired and wireless communication (such as PLC and wireless) [17–19] is a more expensive solution.

Particular attention should be paid to the Power Line Communication systems that communicate on the 230 VAC/50 Hz and 110 VAC/ 50–60Hz power lines, which can be easily adapted—each element that is connected to the power supply can simultaneously be connected to an intelligent building system [20,21].

Depending on the end application, broadband PLC [22] and low-frequency narrowband PLC will be discussed. Broadband PLC (also BPL—Broadband over Power Line) permits a >100 Mbps data rate and usually works on the 2–30 MHz band (also with the possibility to use 30–86 MHz as an additional bandwidth). BPL is usually used in applications for which a relatively high data rate (such as for PLC capabilities) is required, for example, for the in-home distribution of IPTV, HDTV, VoIP, internet content or for communication between electrical systems and appliances. The BPL standards usually define a protocol-dependent MAC layer. Advanced encryption such as a 128-bit AES (Advanced Encryption Standard) encryption for HomePlug AV can also be included. Although the end applications that use BPL offer a relatively high data rate, there are a few limitations such as a limited effective range, an unpredictable response time, an expensive large filter and matching circuits and relatively high processing needs (process encryption, PHY, MAC and IPv4/IPv6 layers). What is more, BPL can potentially cause electromagnetic interference problems in situations in which some of the energy is radiated as radio frequency interference. As for low-frequency narrowband PLCs, according to European CENELEC standard, few frequency bands have been defined for power line communication use. Frequency band A (9 to 95 kHz) is limited to use by energy providers. Although frequency band B (95 to 125 kHz) and frequency band D (140 to 148.5 kHz) are open for end-user applications, no access protocol has been defined. Frequency band C (125 to 140 kHz) is also open for end-user applications but CSMA/CA (Carrier Sense Multiple Access/Collision Avoidance) [23] protocol for the 132.5 kHz frequency has been defined. PRIME and G3 specifications, which are suitable for smart-metering applications [24] and have defined and structured PHY and MAC layers, should also be mentioned. There are also transceivers that are fully configurable. For a low-frequency narrowband PLC that has a scarifying data rate (up to 128 kbps) but has a large effective range (compared to BPL) and a fast response time, its filters and matching circuits are less expensive, and its processing needs are significantly less demanding—simple encryption and the possibility of defining its own MAC layer. The end application was defined as a power line communication home automation system in which short frames (commands) are exchanged between the nodes and for which the effective range and fast system response time is critical. Autonomous nodes are distributed in the system, and therefore a small form factor and low power consumption must be guaranteed. Inexpensive nodes and accompanying elements are also desirable. Therefore, a low-frequency narrowband power line communication opened for end-user applications with defined medium access control and advantageously without any protocol-dependent MAC was selected as one of the possible application scenarios.

In this case, a ready-made STEVAL-IHP005V1 development board with an ST7540 was used to perform the PLC communication tests. The line coupling interface was modified to allow the ST7540 device to transmit and receive on the AC mains line using the available FSK and PSK [25,26] modes within the European CENELEC EN50065-1 standard C band, which is specified for home network systems with the mandatory access protocol CSMA/CA. Customized software was developed for the STM32 microcontroller, which was included in the reference design in order to make it more flexible and suitable for use as a standalone smart PLC node [27].

Currently, many electric energy receivers are strongly non-linear. Unfortunately, these types of receivers often add disturbances to the power supply network in the frequency band that is used for PLC communication [28]. Under certain circumstances, these may interfere with system communication, thereby leading to delays in executing a command. In extremely unfavorable cases, they can prevent communication completely. One of the ways to avoid such problems with a connection is to create an appropriate communication protocol using the appropriate modulation for the signal transmission [29–32]. Unfortunately, in many cases [33], the application of even the most sophisticated communication protocol does not solve the problems connected with the transmission of a useful signal. An additional element of such a system operating using PLC technology should be a filter that separates the undesirable signals (noise, interference, disturbances, signals from other PLC network) from the desired signal in the transmission medium. There are many EMI filters [34,35] on the market, but only few fulfill PLC specific requirements. Usually, PLC EMI filters are relatively large and expensive three-phase filters that are intended to be placed in an electrical switchboard next to the main electricity switch. The cost and form factor of the available PLC EMI filters means that they are not suitable for distributed PLC home automation systems. What is more, in the event that the disturbance source is located after the EMI filter, the communication frequencies are vulnerable to disturbances. An example of such a disturbance source that is located inside the PLC network might be the LED power supply or a fluorescent lamp that is controlled by the home automation system.

There is a need for a compact, single-phase, low-current (5 A), cost-effective filter for the distributed PLC home automation applications that work in CENELEC band C. The main goal is to filter a disturbance source inside the home grid as well as to filter a disturbance source that is located near the communication node.

2. Filter Design

From the point of view of the application of a given electronic system, the basic task of filters is to suppress the undesirable component frequencies that may occur in the control signal. Filter systems are divided into different groups by considering the appropriate criteria. One of the most important of these divisions is the frequency band that is to be suppressed by the filter. Thus, there are different filters—low-pass, high-pass, band-pass including broadband and narrowband (selective) and band-stop—that suppress the signals in a specific frequency band [36]. Other criteria that are considered in the classification of the filters are, for example, the shape of the frequency characteristics: the amplitude and phase, the type of elements that are used and the technology in the system execution. Another important feature of a filter is its level. Filters at I, II and higher levels are used when the higher the filter level, the steeper the edges at the ends of the frequency response are and the frequency response (amplitude) is ideal (rectangular).

The article will focus on the design of an LC filter with chokes in the longitudinal branch and capacitors in the transverse branch as is shown with a diagram in Figure 1a. The role of the inductors is to increase the impedance for a high frequency differential mode component (current or voltage) and the capacitors constitute a low impedance path for differential disturbances at Terminals B_1 and B_2. In addition, the LC parameters were selected so that the resonance frequency was smaller than the 133 kHz communication frequency that had been selected for the purposes of the study and were of an inductance L = 66 μH and a capacity C = 200 nF, respectively. The influence of the change of the inductor parameters was also taken into account [37–39]. The impedance characteristics as seen from Terminals A_1, A_2 and Terminals B_1, B_2 are shown in Figure 2. For the communication frequency f_{com} = 133 kHz, the impedance module was Z_{AA} = 49.17 Ω and Z_{BB} = 6.71 Ω. Such impedance values effectively contribute to the filtering of high frequency disturbances in the communication band when the source of the disturbances is connected to Terminals B_1, B_2.

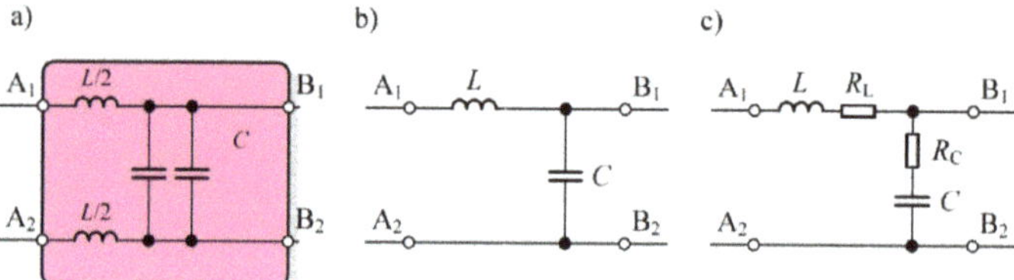

Figure 1. Scheme of the LC single-phase filter with two 33 μH inductor chokes and two 100 nF capacitors: (**a**) scheme, (**b**) a substitute scheme of the ideal filter, (**c**) a filter scheme with the resistive parasitic elements indicated.

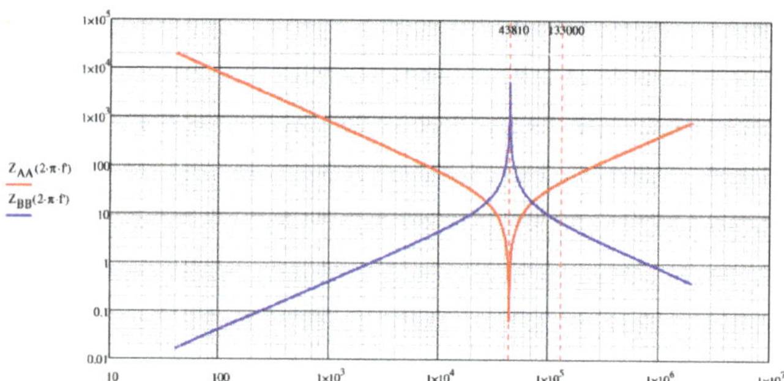

Figure 2. Magnitudes of the impedances seen from terminals A_1, A_2, Z_{AA} and B_1, B_2, Z_{BB} given as the functions of the frequency for the designed filter.

The paper will focus on an analysis of an LC filter with a structure as is shown in Figure 1b. Figure 1c presents parasitic resistance occurring in L and C. These values are dependent on the components and magnetic materials being used [38]. The solution uses two chokes with the toroidal core DTMSS-20/0.033/8.0-V with the inductance L_1 = 33 μH (which gives inductance L = 66 μH) with an allowable current of 8 A and a resistance for the direct current of 15.9 mΩ (total R_{Ldc} = 31.8 mΩ). Two WIMA MKS 100 nF/630 V capacitors with a capacity of 100 nF were used. The filter schemes with a view of the printed circuit are shown in Figure 3.

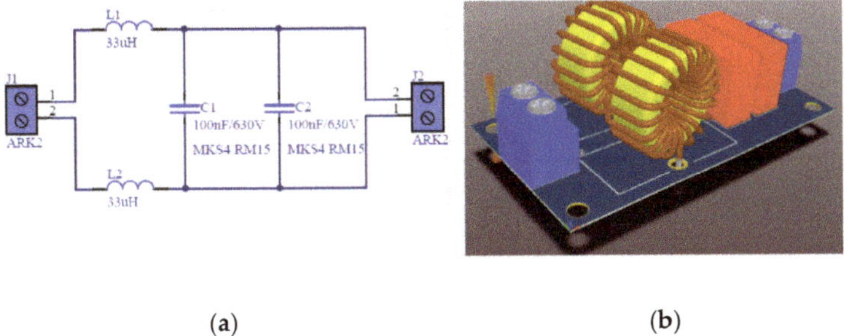

(a)   (b)

Figure 3. LC filter with two THT 33 µH chokes: (**a**) filter scheme, (**b**) 3D view of the board (57.5 mm width and 31.0 mm height, nominal current 5 A).

3. Insertion Loss Filter

In order to determine the insertion loss of the topology LC the filter in Figure 1, filter impedance was measured as a function of the frequency using an Agilent 4294 A Impedance Analyzer in a frequency range of 40 Hz to 2 MHz. Figure 1c shows the parasitic resistances of the reactors and capacitors, which, due to the impedance analyzer measurements, led to the determination of the actual parameters as a function of the frequency. The measurements were taken for Figure 3 of the filter. The filters for the SMD inductors were not analyzed because they had a different substitute pattern in which significant self-capacitances of the inductor winding were revealed. Using the toroid core revealed a relatively low self-capacitance of the inductor winding, and therefore, these inductors were taken into consideration for the filter design. Figure 4 presents the frequency characteristics of inductance L and resistance R_L of the inductors as well as the capacitance C and resistance R_C of the capacitor. The inductance of the inductor was practically constant as a function of frequency, but the capacity of the capacitor changed significantly.

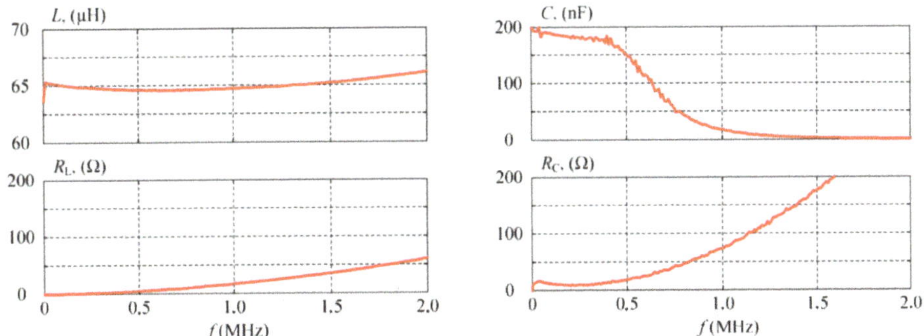

Figure 4. Measured inductance, resistance of the filter inductor, capacitance and its series equivalent resistance of the filter capacitor.

According to the LC filter equivalent parameters, the insertion loss was determined from Terminals A_1, A_2 and from Terminals B_1, B_2. Both insertion losses were determined analytically for circuits as is shown in Figure 5. The insertion loss was defined as in Equation (1)

$$AT(dB) = 20\log\left|\frac{V_1}{V_2}\right|. \tag{1}$$

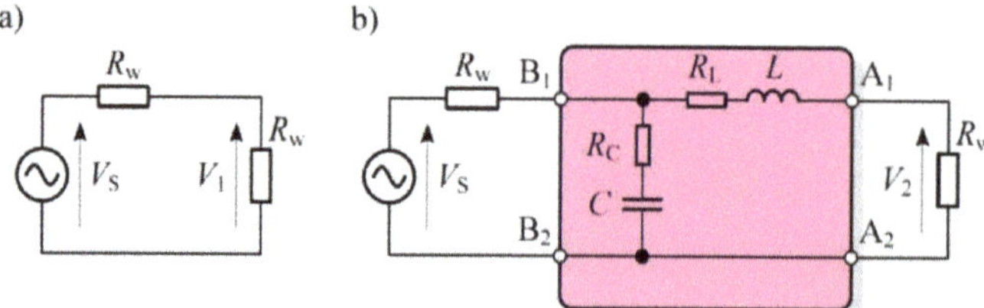

Figure 5. Schematics (circuits) in which the V_1 and V_2 voltages were obtained in order to calculate the insertion loss of the filter from Terminals B to A: (**a**) system without a filter—determination of the voltage V_1 and (**b**) system with a filter—determination of the voltage V_2.

Voltage V_1 is the voltage that occurs on the standard resistor $R_w = 50\ \Omega$ in a case in which there is no filter in the system (Figure 6a); and V_2 is the output voltage on a standard resistor in a case in which the analyzed LC filter is connected between the R_w resistors (Figure 6b).

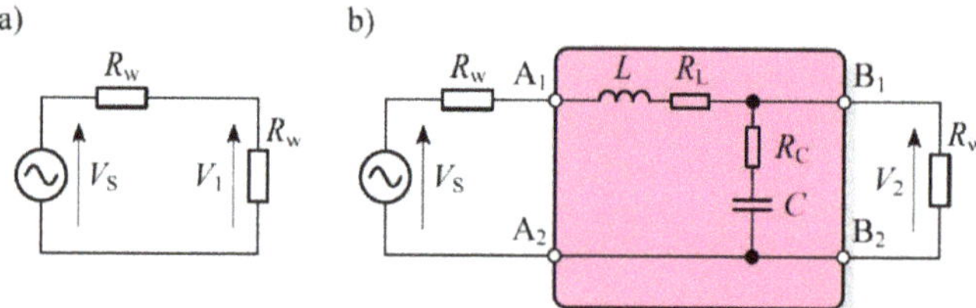

Figure 6. Circuit schemes for which the V_1 and V_2 voltages were required to determine the filter insertion loss from Terminals A to B: (**a**) system without a filter to determine voltage V_1 and (**b**) system with a filter to determine voltage V_2.

According to Figure 6, the following dependences on the voltages V_1 and V_2 can be determined as follows:

$$V_1 = \frac{V_S}{2}, \tag{2}$$

$$V_2 = \left(\frac{V_S \frac{Z_C R_w}{Z_C + R_w}}{R_w + Z_L + \frac{Z_C R_w}{Z_C + R_w}} \right), \tag{3}$$

where $Z_C = \frac{1}{j\omega C} + R_C$; $Z_L = j\omega L + R_L$, thus Equation (3) can be rewritten into Equation (4):

$$AT_{AB}(dB) = 20\log \left| \frac{\frac{V_S}{2}}{\frac{V_S \frac{Z_C R_w}{Z_C + R_w}}{R_w + Z_L + \frac{Z_C R_w}{Z_C + R_w}}} \right| = 20\log \left| \frac{1}{2 \frac{\frac{Z_C R_w}{Z_C + R_w}}{R_w + Z_L + \frac{Z_C R_w}{Z_C + R_w}}} \right|. \tag{4}$$

For the insertion loss that was measured from terminals B to A, the scheme is shown in Figure 5 applies.

Voltage V_1 was the same as in the previous case $V_1 = V_S/2$. Voltage V_2 was determined according to Equation (5) and the filter insertion loss is given as Equation (6).

$$V_2 = \frac{V_S \frac{Z_{LCR}}{Z_{LCR} + R_w}}{R_w + Z_L} R_w, \tag{5}$$

where $Z_{LCR} = \frac{(Z_L+R_w)Z_C}{Z_L+R_w+Z_C}$ thus the filter insertion loss is given as

$$AT_{AB}(dB) = 20\log\left|\frac{\frac{V_S}{2}}{\frac{V_S\left(\frac{Z_{LCR}}{Z_{LCR}+R_w}\right)}{R_w+Z_L}R_w}\right| = 20\log\left|\frac{1}{2\frac{\left(\frac{Z_{LCR}}{Z_{LCR}+R_w}\right)}{R_w+Z_L}R_w}\right|. \tag{6}$$

Equations (4) and (6) are simplified to the same form, which means that the insertion loss of the filter on the side of Terminals A and B are equal and can be described by Equation (7). The characteristics of the insertion loss are presented as the frequency function in Figure 7.

$$AT_{AB}(dB) = AT_{BA}(dB) = 20\log\left|\frac{1}{2\frac{Z_C R_w}{R_w(Z_C+Z_L+R_w)+Z_C(Z_L+R_w)}}\right|. \tag{7}$$

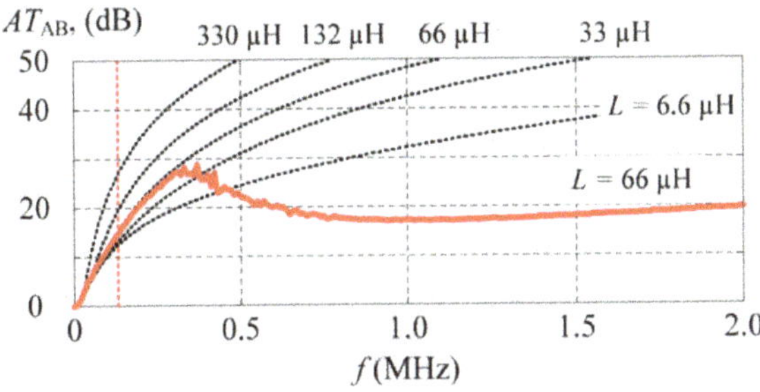

Figure 7. Insertion loss characteristics of the experimental LC filter with the ideal characteristics given for different inductance values and a fixed capacitance of C = 200 nF.

As can be seen in Figure 8, the insertion loss for the 133 kHz frequency that was determined was about 16 dB. Figure 7 shows the characteristics of the filter insertion loss along with the insertion loss of the lossless filters with different inductances and capacitances as a function of the frequency. The insertion loss characteristics are presented for the different inductance values assuming that the rated value was L = 66 µH and the characteristics are given for 0.1L, 0.5L, L, 2L, 5L. For the ideal filters, insertion loss values at f = 133 kHz were 12.5, 13.3, 15.4, 19.7 and 27.1 dB.

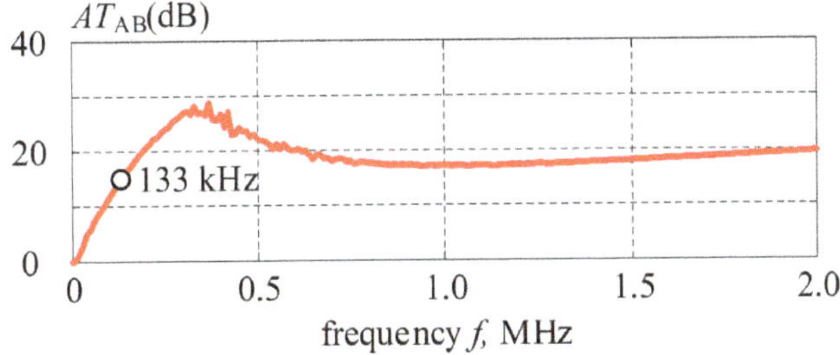

Figure 8. Insertion loss of the LC filter with two 33 µH inductions.

Figure 9 shows the filter insertion loss characteristics of the LC filter together with the ideal characteristics that were obtained for filters with different capacitance values with the assumption that C = 200 nF for 0.1C, 0.5C, C, 2C and 5C. With such insertion loss characteristics for the frequency f = 133 kHz, they were 0.9, 9.3, 15.4, 21.6 and 29.7 dB. Based on the actual characteristics, it can be seen that this solution is ideal for frequencies up to 350 kHz. The reason for the discrepancy between the actual insertion loss characteristics and the ideal characteristics was a decrease in the capacitor capacitance for the frequencies above 350 kHz (Figure 4), while the inductance of the chokes in the frequency range from 40 Hz to 2 MHz remained practically constant. The developed filter is a differential filter and the insertion loss for this filter will never be high as is the case with a common filter in which the common currents component are small.

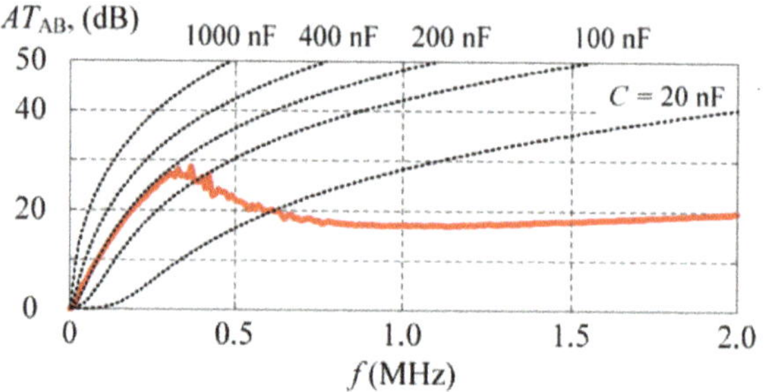

Figure 9. Insertion loss characteristics of the experimental LC filter with the ideal characteristics given for different values of the capacitances and a fixed inductance of L = 66 μH.

4. Thermal Testing of the Filter

If the designed filter is applied next to the disturbance source, it might be exposed to difficult operating conditions such as a constant maximum current flow and poor air circulation. Therefore, heat tests were carried out at a rated current equal to 5 A. The solutions used elements whose rated currents exceeded a 5 A r.m.s. value. The tests were carried out with a load consisting of a set of power supplies with LED lighting that were connected in parallel with a fluorescent lamp and additionally with a thermoregulator with a thyristor voltage controller with an applied phase control. The sum of the currents of the load for the tests was set at 5.4 A. The measurements were performed at an ambient temperature of 19.3 °C. The temperature was measured using a ThermoGear G-30 InfReC thermal imaging camera and a universal meter with an APPA-305 temperature measurement. In each case, the temperature sensor was attached to the surfaces. Therefore, the value from the meter with the temperature sensor was taken as the actual value. A thermal camera was used to determine the points with highest temperature as well as the temperature distribution. In the case of a filter with toroidal throttles, the temperature at a current of 5.4 A in the steady state was 44 °C. The photograph from the thermal camera with the toroidal choke (Figure 10) for the steady state after heating the filter elements with the 5.4 A current indicated the temperature distribution on the individual elements. From this, it can be seen that the highest operating temperatures in the filter were observed in the inductors because they conducted the entire load current. However, their operating temperature remained relatively low.

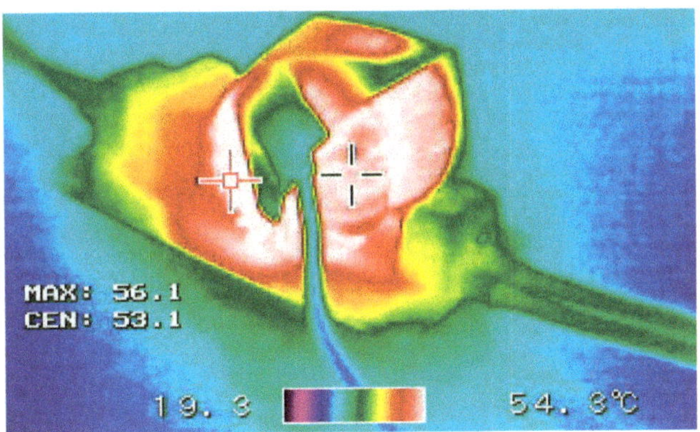

Figure 10. Thermographic photograph of the filter.

5. Verification of the Correct Operation of the Developed Filter

The measurements were carried in five variants. The schematic of the measurement system is shown in Figures 11–15 where: the AC source symbol together with yellow noise symbol represents real-life power grid connection, F—the designed filter, EMI—Schafner interference attenuation filter EMI FN 3256 [40], T_{PLC}—transmitter (switch) of the PLC communication system, R_{PLC}—communication system executive (receiver) and ChX—probe connection voltage differentials to the respective oscilloscope channels. The voltage spectra were measured using the FFT function. The scale of X: 50 kHz/div and Y: 10 dB/div was set and the offset was X: 0 Hz and Y: -30 dB. The load consisted of:

- At the end of the supply line: 3×12 V/100 W halogen power supplies (each one) and 2×12 V/50 W halogen bulbs (each one);
- In the middle of the power line: $1 \times$ fluorescent lamp 2×36 W with an electronic ballast.

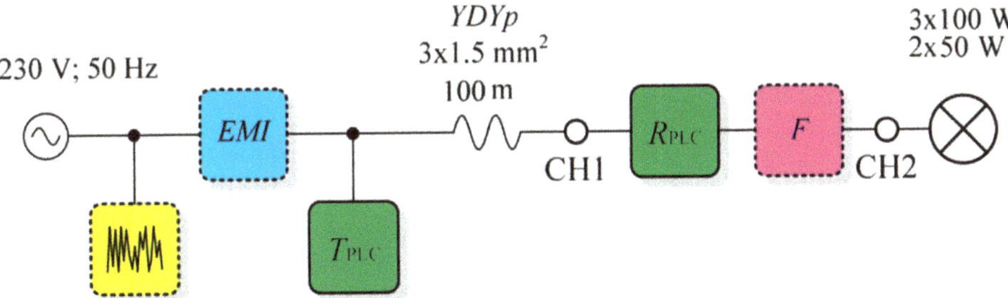

Figure 11. Connection schematic of the model with an EMI filter and without an additional receiver connected in the middle of the line—Variant 1.

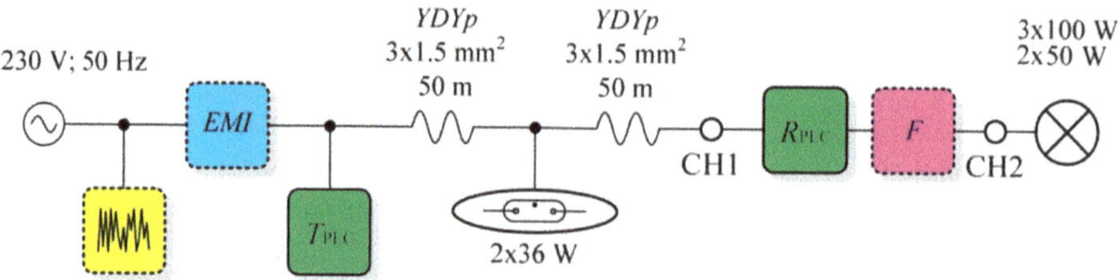

Figure 12. Connection schematic of the model with an EMI filter and an additional receiver connected in the middle of the line (without an F filter)—Variant 2.

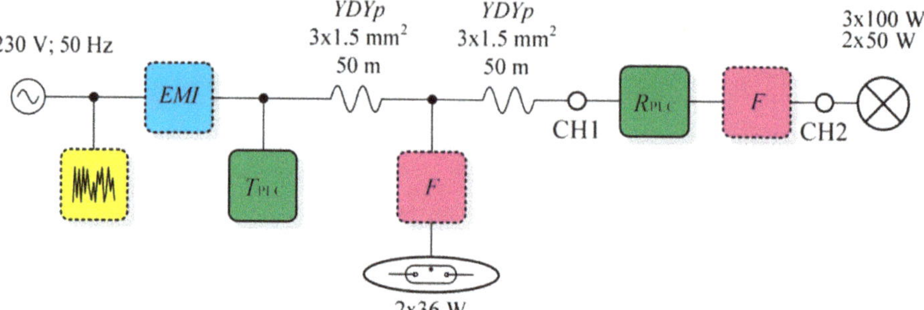

Figure 13. Connection schematic of the model with an EMI filter and an additional receiver connected in the middle of the line by an F filter—Variant 3.

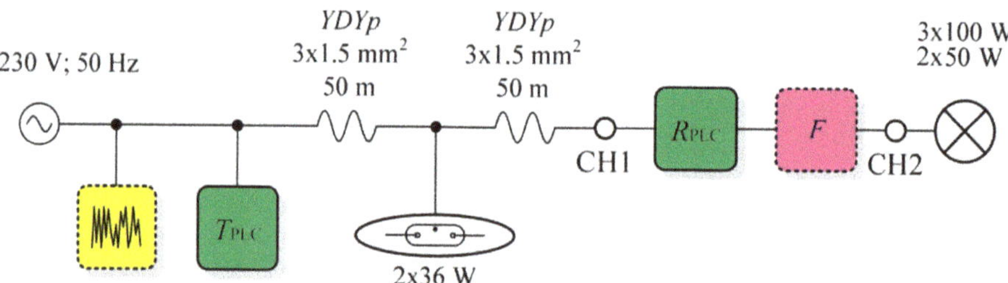

Figure 14. Connection schematic of the model without an EMI filter and with an additional receiver connected in the middle of the line (without the F filter)—Variant 4.

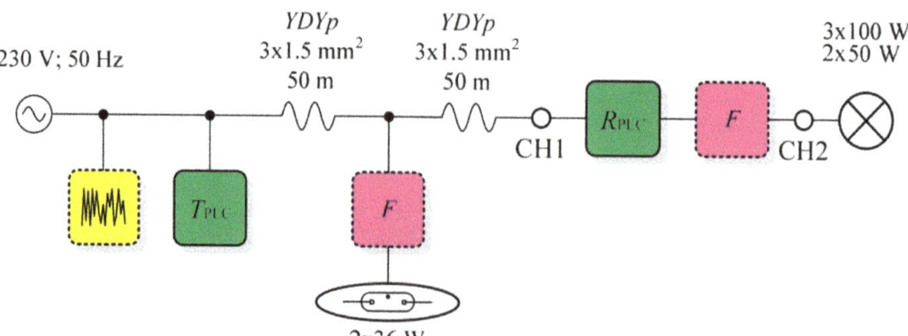

Figure 15. Connection schematic of the model without an EMI filter and with an additional receiver connected in the middle of the line by the F filter—Variant 5.

Additional tests without EMI filters were carried out. The schematic of the measurement system is shown in Figures 14 and 15. Voltage waveforms and voltage harmonics spectra for the two variants configurations is shown in Figure 17.

Based on the results of the measurements (Figures 16 and 17 and Tables 1 and 2), it was found that:

- non-linear type receivers and power supplies weakened the communication signal of the developed system significantly due to the capacitive filter in the input circuits of the power supplies;
- the developed filter significantly improved the communication signal level of the developed system in each of the considered cases;
- comparing the results for f_1 and the measurements for the variant at U_{CH1} and U_{CH2}, the difference in the signal levels at the input and output of the filter was about 5 dBV; however, it must be added that it is difficult to interpret this difference as an insertion loss because sources of interference were present on both sides of the filter;
- by comparing the results for f_2 and the measurements for the variant at U_{CH1} and U_{CH2}, the difference in the signal levels was about 20 dBV;
- adding a fluorescent luminaire in the middle of the model circuit line introduced additional disturbances; these disturbances occurred in the immediate vicinity of 133 kHz, which is used for the transmissions in the developed system;
- it is reasonable to also use the filter for devices that are not controlled by a home automation system that has all types of power supplies installed in the immediate vicinity of the operating system; however, such receivers introduce disturbances that may interfere with the transmissions of the developed system;
- by comparing the relevant cases, a significant reduction of interference at the level of 5 dBV and an improvement of the signal with a transmission frequency of 133 kHz at the level of 20 dBV can be observed, which proves the need to use correctly designed filters for communication in a PLC;
- the use of an EMI filter at the input of the system also resulted in a very strong insertion loss of the communication signal of the developed system, due to the large capacity of the EMI filter;
- obtained attenuation level is sufficient to filter both disturbances and communication signals around 133 kHz;
- with specific filter applications, an independent communication subnetwork can be created (before and after designed filter), such application might be advantageous for PLC home automation system throughput and response time;

Table 1. List of the signal level measurements for the frequency f_2 = 133 kHz and for the highest disturbance components f_1.

Number	Variant	f_1 [kHz]	U_{f1} [dBV]	f_2 [kHz]	U_{f2} [dBV]
1.	1a/U_{CH1}	73	−20.0	133	−11.9
2.	1b/U_{CH2}	73	−15.6	133	−35.0
4.	2c/U_{CH1}	73	−21.9	133	−21.3
5.	2d/U_{CH2}	73	−14.4	133	−39.1
7.	3e/U_{CH1}	83	−18.8	133	−11.9
8.	3f/U_{CH2}	76	−13.1	133	−34.4

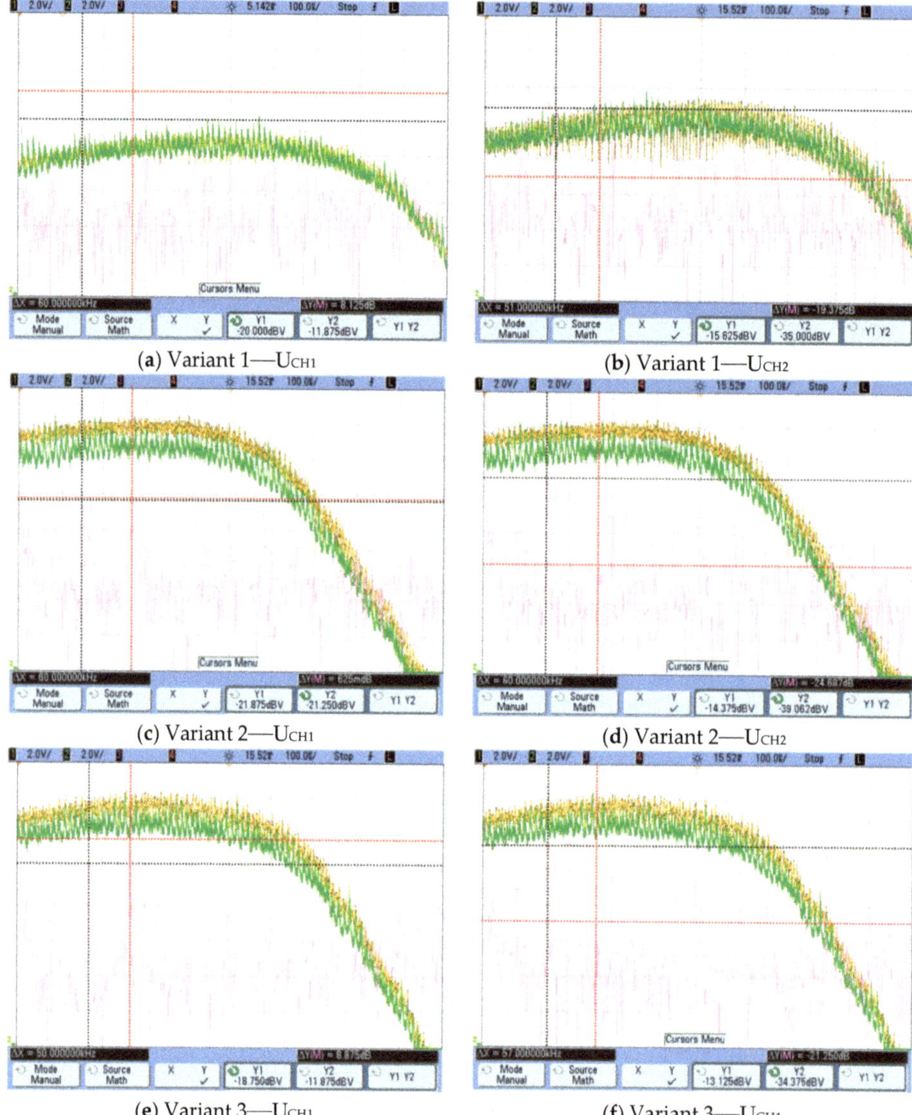

(**a**) Variant 1—U_{CH1}

(**b**) Variant 1—U_{CH2}

(**c**) Variant 2—U_{CH1}

(**d**) Variant 2—U_{CH2}

(**e**) Variant 3—U_{CH1}

(**f**) Variant 3—U_{CH1}

Figure 16. Voltage waveforms and voltage harmonics spectra for the various configurations of the system with the developed filter (X: 50 kHz/div and Y: 10 dB/div). System status: T_{PLC}: broadcast ON; R_{PLC}: ON, measured/analyzed voltages U_{CH1} and U_{CH2}; measured frequency signal level f_1 is the intersection of the black lines and the measured frequency signal level f_2 is the intersection of the red lines.

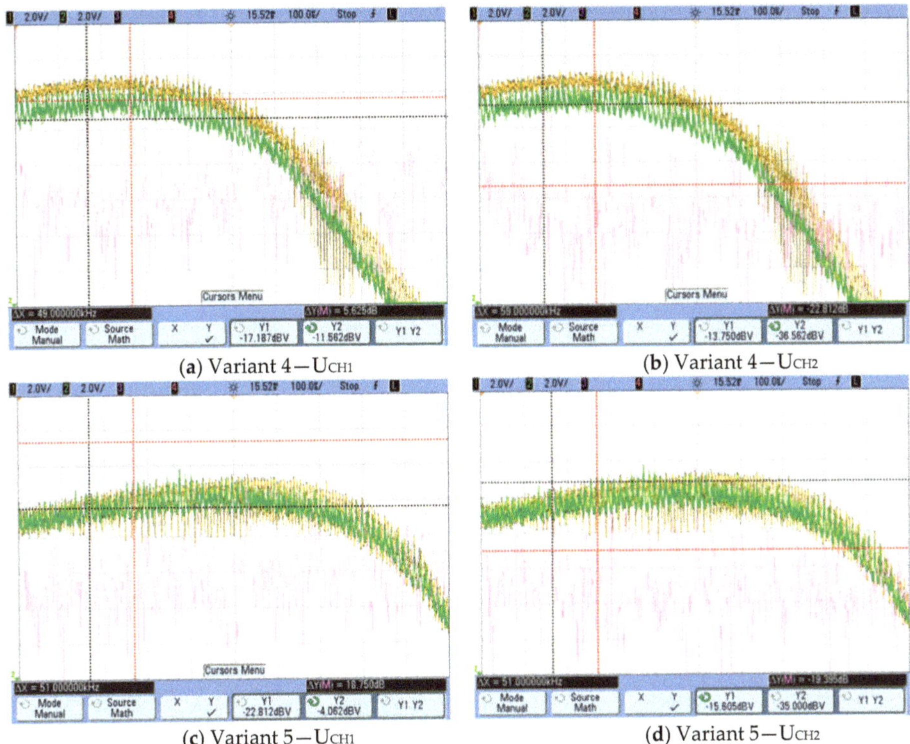

(**a**) Variant 4—U_{CH1}

(**b**) Variant 4—U_{CH2}

(**c**) Variant 5—U_{CH1}

(**d**) Variant 5—U_{CH2}

Figure 17. Voltage waveforms and voltage harmonics spectra for the various configurations of the system with the developed filter (X: 50 kHz/div and Y: 10 dB/div). System status: T_{PLC}: broadcast ON; R_{PLC}: ON, measured/analyzed voltages U_{CH1} and U_{CH2}; measured frequency signal level f_1 is the intersection of the black lines and the measured frequency signal level f_2 is the intersection of the red lines.

Table 2. List of the signal level measurements for the frequency $f_2 = 133$ kHz and for the highest disturbance components f_1.

Number	Variant	f_1 [kHz]	U_{f1} [dBV]	f_2 [kHz]	U_{f2} [dBV]
1.	4a/U_{CH1}	84	−17.2	133	−11.6
2.	4b/U_{CH2}	74	−13.8	133	−36.5
3.	5c/U_{CH1}	82	−22.8	133	−4.1
4.	5d/U_{CH2}	82	−15.6	133	−35.0

6. Statistical Tests Used to Verify the Correct Operation of the Filter in a Model System

In order to verify the correct operation of the filters in the model system, a series of statistical tests were performed with different topologies of the model system. Nonetheless, the topology was always based on the layout shown in Figure 13. In each case, the same board, ST7540 configuration, software, 20-byte frame and transmission scheme were used. The transmission scheme was a series of 50 switches ON and 50 switches OFF commands that were sent with one second delay in between transmissions. A carrier frequency of 132.5 kHz was used with a 2.4 kHz frequency deviation (131.348 kHz for "1" and 133.626 kHz for "0") and 2400 Baud. To improve the robustness, Manchester coding was used. The Forward Error Correction (FEC) technique was used as the error correction mechanism. Situations of correct switching without the need for error correction (CS), correct switching after FEC correction (with FEC) and unsuccessful switching (US) were observed. The series of tests was performed for the following topologies of the model system:

- TEST1—an EMI filter was connected at the model system input and a fluorescent lamp with a power supply system was connected in the middle of the wires, while LED lighting systems with power supply units, a fluorescent lamp and LED lighting were connected behind the proposed filter system (Figure 13).
- TEST2—an EMI filter was present at the input of the model system and a fluorescent lamp with a power supply system was connected in the middle of the wires, LED lighting systems with power supplies worked as the receiver, a fluorescent lamp with no filter was connected and LED lighting was connected with the proposed filtering system (Figure 12).
- TEST3—an EMI filter was present at the input of the model system, while LED lighting systems with power supplies and a fluorescent lamp with a power supply system worked as the receiver and the receivers were connected with the proposed filtering system (Figure 18).
- TEST4—an EMI filter was present at the input of the system, while the receiver was used to work with the LED lighting systems with the power supplies and a fluorescent lamp with a power supply system and the receivers operated without the proposed filtering system (Figure 19).

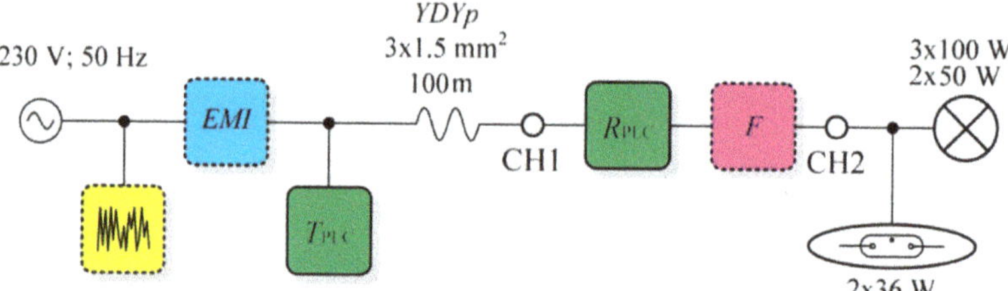

Figure 18. Connection schematic of the model with an EMI filter while LED lighting systems with power supplies and a fluorescent lamp with a power supply system working as the receiver, the receivers were connected with the proposed filtering system (TEST 3).

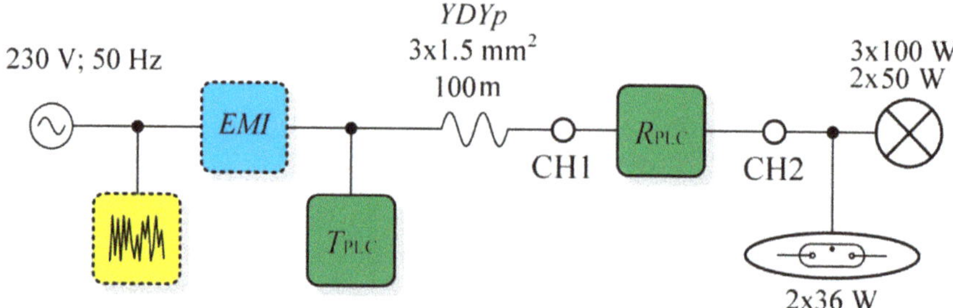

Figure 19. Connection schematic of the model with an EMI filter while LED lighting systems with power supplies and a fluorescent lamp with a power supply system working as the receiver, the receivers were connected without the proposed filtering system (TEST 4).

Based on the statistical tests that were performed (Table 3), the following can be concluded:

- in the case of the switch OFF tests, there were more problems with the transmission and cases in which there was no response to the signal from the transmitter more often. This was because the receivers generated interference that hindered the reception of the signals;

- the absence of additional filters in the system made it practically impossible to control the receivers when connecting in the middle of wires or as a controlled receiver of the fluorescent lamp (TEST2 and TEST4);
- in the case of an EMI filter and the use of the developed filters, full system efficiency was obtained from the point of view of controlling the receivers.

Table 3. Results of the statistical tests, correct switching (CS), unsuccessful switching (US) and correct switching with Forward Error Correction (FEC).

Number	Setting the Model	Switch ON			Switch OFF		
		CS	With FEC	US	CS	With FEC	US
1.	TEST1	30	20	0	35	15	0
2.	TEST2	0	2	48	0	2	48
3.	TEST3	26	24	0	30	20	0
4.	TEST4	0	50	0	0	2	48

7. Conclusions

Today, PLC communication is being used more and more to implement home automation solutions as well as in intelligent homes and remote controls for electronic equipment. In the standard solutions, corrections resulting from error transmissions are based on complex digital modulation methods and algorithms for validating the transmitted data without paying attention to the origin of the errors. In this article, we have focused on the implementation of a filtering system for interference in the 120–150 kHz band that is introduced into the network by receivers. Such a filter separates the desired signal from the interference that occurs in a network, which can result in communication errors. An additional advantage of the developed solution is the ability to provide physical signal separation between two PLC sub-systems (before and after the filter system). This approach might possibly make the response faster and increase the throughput of PLC sub-systems. The entire project cycle was presented beginning with specifying the requirements, through the design, simulation, execution and testing under operating conditions. All of the tests were carried out in real life working conditions with an additional load and the validity and impact of using the EMI filter on the efficiency of information transmissions in the PLC was considered. Through these efforts, it was shown that by using an appropriate filtering system, it is possible to use less complex methods of error correction in the transmission of signals, thus enabling an increase in the transmission speed. In addition, due to the reduction in the complexity of the corrective methods, it is possible to use the computed savings in the capacities to implement other important functionalities or to improve the transmission security without affecting the transmission delay and user experience.

Author Contributions: Conceptualization, K.B. and D.W.; Methodology, D.W., K.B. and M.Z.; software, D.W. and K.B.; Validation, D.W., K.B., M.Z., A.L., J.M.; Formal analysis, D.W. and K.B.; Investigation, K.B., D.W.; Resources, D.W., K.B. and M.Z., Writing—original draft preparation, K.B.; Writing—review and editing D.W., K.B. and Z.R.; Visualization, K.B. and D.W.; Supervision, Z.R.; Funding acquisition, K.B.

Funding: This research was partially supported by the Polish Ministry of Science and Higher Education funding for statutory activities (BK, BKM) of the Institute of Electronics, Silesian University of Technology.

Acknowledgments: The calculations were performed using the IT infrastructure that was funded by the GeCONiI project (POIG.02.03.01-24-099/13).

Conflicts of Interest: The authors declare no conflict of interest.

References

1. Schwartz, M. Carrier-wave telephony over power lines: Early history. *IEEE Commun. Mag.* **2009**, *47*, 14–18. [CrossRef]
2. Whiffen, T.; Naylor, S.; Hill, J.; Smith, L.; Callan, P.; Gillott, M.; Wood, C.; Riffat, S. A concept review of power line communication in building energy management systems for the small to medium sized non-domestic built environment. *Renew. Sustain. Energy Rev.* **2016**, *64*, 618–633. [CrossRef]
3. Alanne, K.; Cao, S. An overview of the concept and technology of ubiquitous energy. *Appl. Energy* **2019**, *238*, 284–302. [CrossRef]
4. Arcia-Garibaldi, G.; Cruz-Romero, P.; Gómez-Expósito, A. Future power transmission: Visions, technologies and challenges. *Renew. Sustain. Energy Rev.* **2018**, *94*, 285–301. [CrossRef]
5. Sendin, A.; Pena, I.; Angueira, P. Strategies for Power Line Communications Smart Metering Network Deployment. *Energies* **2014**, *7*, 2377–2420. [CrossRef]
6. Sharma, K.; Saini, L.M. Power-line communications for smart grid: Progress, challenges, opportunities and status. *Renew. Sustain. Energy Rev.* **2017**, *67*, 704–751. [CrossRef]
7. Available online: http://www.marketresearchstore.com/report/global-building-automation-systems-market-industry-analysis-73569 (accessed on 20 March 2019).
8. Ikpehai, A.; Adebisi, B.; Rabie, K.; Haggar, R.; Baker, M. Experimental Study of 6LoPLC for Home Energy Management Systems. *Energies* **2016**, *9*, 1046. [CrossRef]
9. Ikpehai, A.; Adebisi, B.; Rabie, K. Broadband PLC for Clustered Advanced Metering Infrastructure (AMI) Architecture. *Energies* **2016**, *9*, 569. [CrossRef]
10. Mlynek, P.; Misurec, J.; Koutny, M. Modeling and evaluation of power line for Smart grid communication. *Przegląd Elektrotechniczny* **2011**, *87*, 228–232.
11. Lesek, F.; Petr, F. A Multiconductor Model of Power Line Communication in Medium-Voltage Lines. *Energies* **2017**, *10*, 816.
12. Galli, S.; Scaglione, A.; Wang, Z. For the grid and through the grid: The role of power line communications in the smart grid. *Proc. IEEE* **2011**, *99*, 998–1027. [CrossRef]
13. Rastegar, M.; Fotuhi-Firuzabada, M.; Lehtonenb, M. Home load management in a residential energy hub. *Electr. Power Syst. Res.* **2015**, *119*, 322–328. [CrossRef]
14. Zhu, Y.; Wu, J.; Wang, R.; Lin, Z.; He, X. Embedding Power Line Communication in Photovoltaic Optimizer by Modulating Data in Power Control Loop. *IEEE Trans. Ind. Electron.* **2019**, *66*, 3948–3958. [CrossRef]
15. Elazhary, H. Internet of Things (IoT), mobile cloud, cloudlet, mobile IoT, IoT cloud, fog, mobile edge and edge emerging computing paradigms: Disambiguation and research directions. *J. Netw. Comput. Appl.* **2019**, *128*, 105–140. [CrossRef]
16. Kamal, M.; Parvin, S.; Saleem, K.; Al-Hamadi, H.; Gawanmeh, A. Efficient low cost supervisory system for Internet of Things enabled smart home. In Proceedings of the IEEE International Conference on Communications Workshops, Paris, France, 21–25 May 2017; pp. 864–869.
17. Li, M.; Lin, H. Design and Implementation of Smart Home Control Systems Based on Wireless Sensor Networks and Power Line Communications. *IEEE Trans. Ind. Electron.* **2015**, *62*, 4430–4442. [CrossRef]
18. Mendes, T.; Godina, R.; Rodrigues, E.; Matias, J.; Catalão, J. Smart home communication technologies and applications: Wireless protocol assessment for home area network resources. *Energies* **2015**, *8*, 7279–7311. [CrossRef]
19. Han, J.; Choi, C.; Park, W.; Lee, I.; Kim, S. Smart home energy management system including renewable energy based on ZigBee and PLC. *IEEE Trans. Consum. Electron.* **2014**, *60*, 198–202. [CrossRef]
20. Fernandes, V.; Poor, H.; Ribeiro, M. Analyses of the Incomplete Low-Bit-Rate Hybrid PLC-Wireless Single-Relay Channel. *IEEE Internet Things J.* **2018**, *5*, 917–929. [CrossRef]
21. Zhao, X.; Zhang, H.; Lu, W.; Li, L. Approach for modelling of broadband low-voltage PLC channels using graph theory. *IET Commun.* **2018**, *12*, 1524–1530. [CrossRef]
22. Pinto-Benel, F.; Cruz-Roldán, F. Exploring the performance of prototype filters for broadband PLC. In Proceedings of the IEEE International Symposium on Power Line Communications and its Applications (ISPLC), Madrid, Spain, 3–5 April 2017; pp. 1–6.
23. Choi, H.; Moon, J.; Lee, I.; Lee, H. Carrier Sense Multiple Access with Collision Resolution. *IEEE Commun. Lett.* **2013**, *17*, 1284–1287. [CrossRef]

24. Lopez, G.; Moreno, J.I.; Amaris, H.; Salazar, F. Paving the road toward Smart Grids through large-scale advanced metering infrastructures. *Electr. Power Syst. Res.* **2015**, *120*, 194–205. [CrossRef]
25. Peck, M.; Alvarez, G.; Coleman, B.; Moradi, H.; Forest, M.; Aalo, M. Modeling and Analysis of Power Line Communications for Application in Smart Grid. In Proceedings of the 15th LACCEI International Multi-Conference for Engineering, Education and Technology, Boca Raton, FL, USA, 19–21 July 2017; pp. 1–6.
26. Cavdar, I.H. Performance analysis of FSK power line communications systems over the time-varying channels: Measurements and modeling. *IEEE Trans. Power Deliv.* **2004**, *19*, 111–117. [CrossRef]
27. Available online: https://www.st.com/resource/en/data_brief/steval-ihp005v1.pdf (accessed on 20 March 2019).
28. Lampe, L.; Rahman, M.; Saffar, H. Characteristics of power line networks: Diversity and interference alignment. In Proceedings of the IEEE International Symposium on Power Line Communications and its Applications (ISPLC), Madrid, Spain, 3–5 April 2017; pp. 1–6.
29. Juwono, F.; Guo, Q.; Chen, Y.; Xu, L.; Huang, D.; Wong, K. Linear Combining of Nonlinear Preprocessors for OFDM-Based Power-Line Communications. *IEEE Trans. Smart Grid* **2016**, *7*, 253–260. [CrossRef]
30. Cui, Y.; Liu, X.; Cao, J.; Xu, D. Network Performance Optimization for Low-Voltage Power Line Communications. *Energies* **2018**, *11*, 1266. [CrossRef]
31. Nombela, F.; García, E.; Ureña, J.; Hernández, A.; Poudereux, P. Robust synchronization algorithm for broadband PLC based on Wavelet-OFDM. In Proceedings of the IEEE 20th Conference on Emerging Technologies & Factory Automation, Luxembourg, 8–11 September 2015; pp. 1–7.
32. Mathur, A.; Bhatnagar, M.; Panigrahi, B. Maximum likelihood decoding of QPSK signal in power line communications over Nakagami-m additive noise. In Proceedings of the IEEE International Symposium on Power Line Communications and its Applications, Austin, TX, USA, 29 March–1 April 2015; pp. 7–12.
33. Okazima, N.; Baba, Y.; Nagaoka, N.; Ametani, A.; Temma, K.; Shimomura, T. Propagation Characteristics of Power Line Communication Signals Along a Power Cable Having Semiconducting Layers. *IEEE Trans. Electromagn. Compat.* **2010**, *52*, 756–769. [CrossRef]
34. Ozenbaugh, R.L. *EMI Filter Design*, 2nd ed.; Marcel Dekker Inc.: New York, NY, USA, 2001.
35. Verma, R.; Maity, T.; Hofsajer, I. Multipath conductors for EMI filter: Recent developments. *Iet Sci. Meas. Technol.* **2018**, *12*, 575–580. [CrossRef]
36. Das, J. Passive filters-potentialities and limitations. *Ieee Trans. Ind. Appl.* **2004**, *40*, 232–241. [CrossRef]
37. Bernacki, K.; Rymarski, Z.; Dyga, Ł. Selecting the coil core powder material for the output filter of a voltage source inverter. *Electron. Lett.* **2017**, *53*, 1068–1069. [CrossRef]
38. Rymarski, Z. Measuring the real parameters of single-phase voltage source inverters for UPS systems. *Int. J. Electron.* **2017**, *104*, 1020–1033. [CrossRef]
39. Rymarski, Z.; Bernacki, K.; Dyga, Ł. The influence of the properties of magnetic materials on a voltage source inverter control. In Proceedings of the IEEE Conference on Control Applications (CCA), Nice/Antibes, France, 8–10 October 2014; pp. 1127–1132.
40. Available online: https://www.schaffner.com/product-storage/datasheets/fn-3256/ (accessed on 20 March 2019).

Article

Design and Development of a Reduced Form-Factor High Accuracy Three-Axis Teslameter

Johann Cassar [1,*], Andrew Sammut [1], Nicholas Sammut [2], Marco Calvi [3], Sasa Dimitrijevic [4] and Radivoje S. Popovic [4]

1   Faculty of Engineering, University of Malta, MSD 2080 Msida, Malta; andrew.sammut@um.edu.mt
2   Faculty of ICT, University of Malta, MSD 2080 Msida, Malta; nicholas.sammut@um.edu.mt
3   Photon Science Division of the Paul Scherrer Institute, 5232 Villigen PSI, Switzerland; marco.calvi@psi.ch
4   Sentronis ad, 18115 Nis, Serbia; sasa_dimitrijevic@sentronis.rs (S.D.); rpopovic@senis.ch (R.S.P.)
*   Correspondence: johann.cassar@um.edu.mt; Tel.: +356-2340-3339

Received: 11 March 2019; Accepted: 22 March 2019; Published: 26 March 2019

Abstract: A novel three-axis teslameter and other similar machines have been designed and developed for SwissFEL at the Paul Scherrer Institute (PSI). The developed instrument will be used for high fidelity characterisation and optimisation of the undulators for the ATHOS soft X-ray beamline. The teslameter incorporates analogue signal conditioning for the three-axes interface to a SENIS Hall probe, an interface to a Heidenhain linear absolute encoder and an on-board high-resolution 24-bit analogue-to-digital conversion. This is in contrast to the old instrumentation setup used, which only comprises the analogue circuitry with digitization being done externally to the instrument. The new instrument fits in a volumetric space of 150 mm × 50 mm × 45 mm, being very compact in size and also compatible with the in-vacuum undulators. This paper describes the design and the development of the different components of the teslameter. Performance results are presented that demonstrate offset fluctuation and drift (0.1–10 Hz) with a standard deviation of 0.78 µT and a broadband noise (10–500 Hz) of 2.05 µT with an acquisition frequency of 2 kHz.

Keywords: analogue-to-digital conversion; ATHOS soft X-ray beamline; broadband noise; Hall probe; offset fluctuation and drift; three-axis teslameter; undulator

---

1. Introduction

Various parameters are typically taken into consideration in the measurement of magnetic fields and primarily include the field range, the bandwidth, the reproducibility and the accuracy [1,2].

Whilst static or quasi DC homogenous magnetic fields are best measured using NMR (Nuclear Magnetic Resonance) instruments with the best absolute accuracies, these have a very limited bandwidth [1]. The choice of silicon-based Hall magnetometers is very attractive when a balance between the precision and the bandwidth response of the instrument is required as is the case in this application. Also, three-axes Hall probes fulfill the necessity in measuring all three components of a magnetic field simultaneously along a straight line as required in undulators.

The measurement with Hall probes is particularly suitable for a broad range of non-homogenous magnetic fields up to 10 T and more. However, the absolute accuracy of the measurement is affected by parasitic effects such as nonlinearity, temperature dependence, planar Hall effect and stability of the offset [1–3]. Hence apart from calibration, these inherent problems in Hall probes require compensation through special biasing and interfacing techniques.

The small and compact volume of the Hall probe allows for high spatial resolution of the magnetic measurement. This makes them very suitable and the preferred choice in mapping the magnetic field in insertion devices (such as undulators) especially when these have a very narrow gap and the bandwidth requirements are not particularly high.

Magnetic field mapping of undulators sets the context of the main application of the development of the three-axis teslameter presented in this paper. In [4], a Hall-probe bench for insertion device characterization at the Brazilian Synchrotron Light Laboratory (LNLS) is presented whereby three single-axis Hall probes are orthogonally assembled and used for the magnetic field component measurements. The use of Hall probes as measuring devices are preferred over rotating coil techniques in this scenario. For such a setup the dimensions of the sensor must be significantly smaller than a single undulator period for proper mapping to be done. This makes Hall probes very advantageous to use over short moving coil techniques. Through these measurements, the existence of random local angular kicks along the undulator axis can also be investigated. Post-processing of the data results in the measurement of the phase error along the undulator axis. On-the-fly feedback scanning is also essential to reduce the sensor vibration during traversal motion. A high level of accuracy in the sensor position on the longitudinal axis is also critical, as pointed out in [2].

Other techniques are used in order to measure the integral field errors along the undulator. These give information on the total change in the angle and position of the beam trajectory at the exit of the undulator. In [5], a method is presented where the field integrals are measured using a multistrand wire stretched inside the undulator. As the wire is moved with constant velocity of translation, the first integral is found. The cross motion of the wire at the undulator ends measures the second field integral.

Therefore, knowledge of these errors is very important for a complete characterization of the magnetic field in an undulator. After being measured, the errors are corrected by magnets shimming as explained in [6].

Following this background on the problem definition, the rest of the paper tackles the development of a novel three-axis teslameter interfaced with a SENIS type-S Hall probe [7]. This instrument will be used with a new magnetic measurement bench developed at the Paul Scherrer Institute (PSI) for the highly precise magnetic field measurements of the ATHOS soft X-ray line [8].

The main motivation of the development of this instrumentation is to improve the current instrumentation used [9] to characterize the ATHOS line of undulators. The main aim is to integrate the full analogue and digital electronics on a single board with a tenfold size reduction and increase the performance and features compared to the old instrumentation. Another goal is to provide an instrument that can easily be used for similar applications in other machines hence it is also important to provide a more cost-effective and maintenance-free setup.

2. Hall Probe Theory

As explained in [10], the resolution of a magnetic sensor depends on its intrinsic noise, offset instability and the magnetic sensitivity. Silicon Hall sensors are typically realized as an n-well plate with four contacts being easily modelled as a Wheatstone bridge [11,12]. Since a current is made to flow through two opposing terminals, a magnetic field perpendicular to the plate makes electrons drift towards one side thereby generating a potential difference known as the Hall voltage.

Due to doping inhomogeneity and variations in the depth of the n-well, the resistance in the various branches of the Wheatstone bridge usually do not match, thus the sensor will exhibit a certain amount of offset when biased as shown in Figure 1. The effect of the offsets can be significantly reduced by employing the spinning current technique [3,11,12]. Swapping the functions of the readout and bias electrodes and thus changing the direction of the bias current through the sensor, swaps the relative polarities of the sensor's offset and Hall voltage. Therefore, doing this periodically, results in the offset being modulated to the spinning frequency while the Hall voltage is recovered by averaging the voltage on the other two contacts.

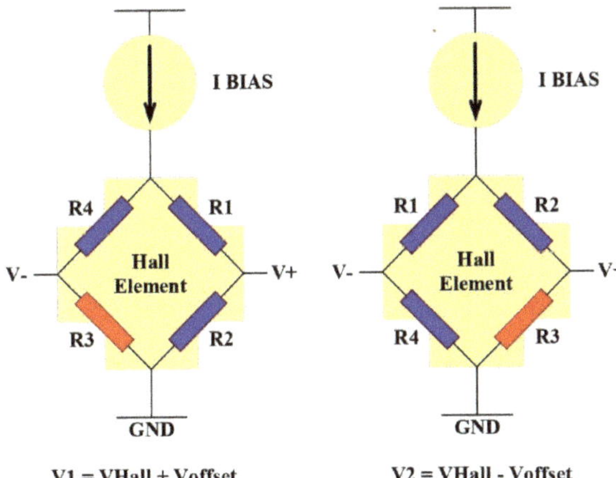

Figure 1. Wheatstone bridge model of a Hall sensor biased with a constant current source and exhibiting an offset voltage.

When exposing the Hall probe to a nonorthogonal magnetic field, the output Hall voltage appears to be the sum of the normal and planar Hall voltages. The planar Hall voltage is dependent on the magnetic field and appears also as an offset voltage where it is proportional to the square of the planar field component. The spinning current method also suppresses the planar Hall Effect [3].

A challenge in the application of the spinning current technique as explained in [10] is the resultant switching spikes that occur upon current reversal. Unfortunately, these spikes result in additional noise, offset drift and $1/f$ noise that must addressed properly in the electronic design since they impact the overall performance of the magnetic measurements. This is the main limitation that makes it difficult to approach the physical limit of the magnetic resolution of the Hall device. High-order low-pass filtering circuitry must be used in order to attenuate sufficiently high order harmonics generated by these spikes. This is tackled in the actual design of the instrument as explained in Section 3.

3. Architecture and Components of the Teslameter

This section provides a detailed overview of the complete architecture of the newly developed teslameter with all details pertaining to the circuitry involved and the interfacing operation capabilities of the instrument. The advancement and the state of the art of this instrument is the incorporation of all the analogue and digital circuitry as explained in this section all on a single printed circuit board (PCB) with very tight dimensions. In addition, this instrument also consists of a digital interface to a Heidenhain linear absolute encoder thus making it possible in providing synchronized position and magnetic field readings. This instrument provides a tailored solution for undulator mapping applications with better performance in both synchronization timing and noise performance, as well as a more complex, flexible and hence precise and accurate calibration routine to be applied.

3.1. Architecture

A block diagram of the instrument architecture is presented in Figure 2. The indicated analogue circuitry consists of identical spinning current and voltage readout circuitry for the interfacing of a three-axis SENIS H3A Hall probe [7]. The analogue differential voltage of each axis is equally amplified using a very low noise instrumentation amplifier and low pass filtered using a third order Butterworth antialiasing filter to the required 500-Hz bandwidth. The signal path is kept fully differential down to the AD converter (ADC). The 24-bit, 4-Ch, simultaneous sampling Delta-Sigma analogue-to-digital converter is the core of the instrument that provides the necessary digitization of the three magnetic field axes and the Hall probe temperature PT100 analogue reading.

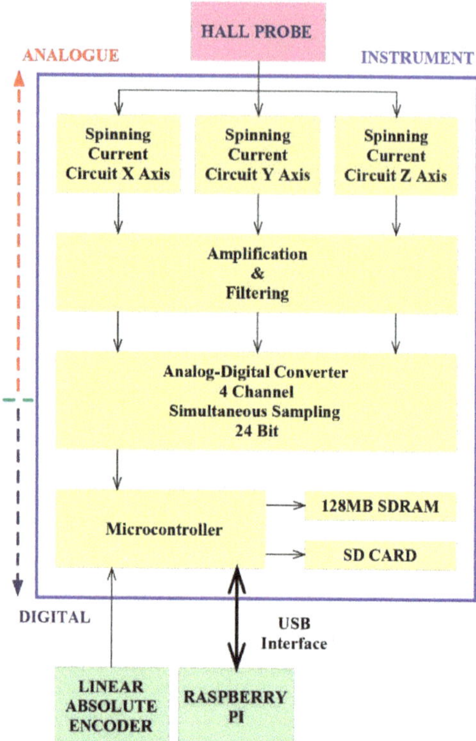

Figure 2. Architecture block diagram of the whole instrument showing the different functions in the analogue and digital sides.

The digital part is organized around the C2000 series Delfino TMS320F28379D microcontroller [13]. This powerful 32-bit floating point and dual core industrial microcontroller is designed for advanced closed-loop control applications and provides 200 MHz of signal processing performance in each core. The microcontroller handles all the communication interfaces with all the peripherals explained further in the next section.

The instrument circuitry board is an eight-layer PCB with physical dimensions of 144 mm in length by 44-mm wide. The eight-layer choice was deemed to be the best option considering the circuitry complexity involved and the physical size restriction that the instrument must fit in. A top view photo of the PCB is presented in Figure 3. The PCB layer stack-up choice incorporates two middle signal layers which are sandwiched between two ground planes for superior electromagnetic compatibility (EMC) performance in comparison to a six-layer stack up choice mainly due to the additional ground plane. The two ground planes enable the incorporation of ground-to-ground vias between the two ground planes near the signals vias in order to provide an adjacent return path for the current.

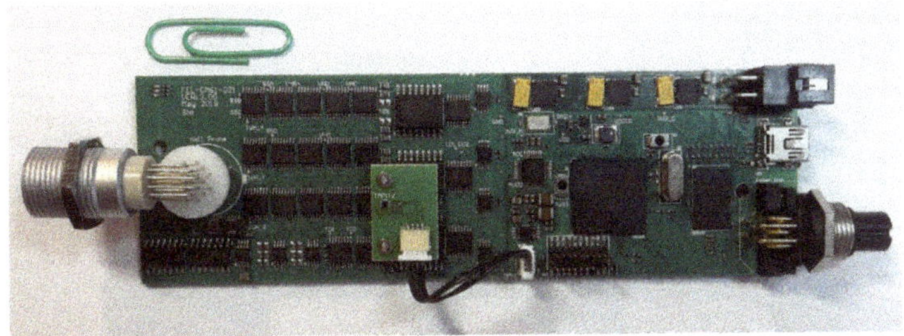

Figure 3. Top view of the eight-layer PCB, the left side comprises the analogue circuitry with the Hall probe connector at the extreme left, the ADC and the microcontroller somewhat in the middle of the board and the power supply connector, USB and encoder connector on the right hand side of the board.

These two signal layers are used mainly for the routing of the very sensitive and noise prone analogue tracks and high speed signal tracks in the digital section of the board. Adjacent to the two ground layers, a power plane layer is included on each side to accommodate the positive and negative power supply rails. The general routing of the board is done through the two outermost top and bottom signal layers. As copper planes are present beneath the outermost signal layers, ground return paths are always present, minimizing crosstalk and distortion between adjacent signal tracks.

In such a mixed signal PCB, proper design considerations are taken for the proper layout of the ground plane. Failure to do so implies a degradation in the noise performance of the instrument as the digital switching noise couples with the sensitive analogue signals. A void slit in the ground plane isolates the analogue precision circuitry and only a single star point beneath the ADC provides the necessary high impedance connection between the analogue and the digital ground planes. This minimizes any unwanted ground return paths from the digital side to enter in the analogue side [14–16].

3.2. Instrument Components

3.2.1. Spinning Current Circuitry

In the spinning current modulation process, the current through the Hall plate is injected alternatively along the north–south or the west–east arms. The current direction and voltage readout from the four-terminal Hall plate is controlled using the ADG1612 CMOS analogue switches. These switches have a nominal on-state resistance of 1 Ω [17]. The flatness of the on-state resistance being defined as the difference between the maximum and minimum values of on-resistance as measured over the specified analogue signal range is 0.2 Ω. As this change in resistance over the range introduces distortion, these switches were chosen for their very flat profile thus ensuring excellent linearity and low distortion. The switching of the CMOS analogue switches are controlled using synchronized pulse width modulation (PWM) control signals from the microcontroller.

Jitter-free operation is achieved through the use of hardware implemented PWM modules within the microcontroller. Once configured these free running PWM modules require little overhead and do not interfere with the operation of the microcontroller in data acquisition. Synchronization of the current direction switching and the voltage readout switching is ensured by setting one of the modules with a master time base and the downstream modules are elected to run in synchronization with the master.

3.2.2. Spinning Output Amplification

An attentive noise analysis study was conducted for the best possible choice of the instrumentation amplifiers which ultimately limits the noise floor of the instrument. The two most important factors that outline the figure of merit for the amplifier's noise voltage and current response are flicker noise, which is mostly dominant at low frequencies, and shot noise being dominant at frequencies beyond the corner frequency. In the analysis presented, the measurements are referred to the amplifier inputs in order to remove the need to account for the amplifier's gain.

This Hall plate has R_{out} = 300 Ω with $4kT.R_{out}$ thermal noise which amounts to 2.222 nV/$\sqrt{\text{Hz}}$. The noise level generated by the amplifier current noise is 90 pV/$\sqrt{\text{Hz}}$ due to the source resistance. Therefore the summation of the noise from the source resistance is 2.223 nV/$\sqrt{\text{Hz}}$. From this it is clear that due to the low source resistance, the voltage noise dominates over the current noise. As the additional noise voltage brought by the op amp is 8 nV/$\sqrt{\text{Hz}}$ at a bandwidth of 1 kHz [18] this results in a total noise spectral density of 8.303 nV/$\sqrt{\text{Hz}}$.

The gain of the amplification stage is set to 4.4 and the bandwidth of the low pass filter is set to 500 Hz with a brick wall correction factor of 1.57. Therefore, the noise voltage root mean square (RMS) value at the output of the amplifier where e_t is the total noise spectral density of 8.303 nV/$\sqrt{\text{Hz}}$ is given by Equation (1):

$$V_t = G \times e_t \times \sqrt{\text{BW}} = 1.023 \times \mu \times V_{rms} \tag{1}$$

Therefore, by combining the different component noise sources both from the sensor and the op amp, the noise floor at the output of the amplification stage is approximated to be $1.023 \times \mu \times V_{rms}$.

3.2.3. Current Source Hall Probe Biasing

The sensitivity response of the Hall element depends on the current magnitude passed through it. Some factors must be considered in choosing this current magnitude, as the lower the current is, the more gain one has to apply in order to achieve the desired full dynamic range. Higher gains generally imply higher noise figures, so a tradeoff was found experimentally in determining the best excitation DC current. However higher currents passed through the Hall element automatically result in higher sensitivity dependence on the probe temperature. A 2.5 mA Howland Current Source circuit [19] is used to drive each axis of the probe. This current source relies on a very high precision 2.5 V reference [20] being buffered to the Howland op amp circuit to avoid any loading and drifts of the reference voltage. One of the main factors that are considered in the design of this current source is the matching of the four resistors attaining the negative feedback of the op amp which is fundamental otherwise a dependence of the output current on the load magnitude occurs. For this reason, a tightly matched resistor array with a very low temperature coefficient was used for each current source.

3.2.4. Interfacing of the Hall Probe PT100

Readout of a voltage proportional to the Hall probe temperature is implemented by passing a constant current through the on chip PT100 whose resistance varies linearly with temperature. This is necessary in applying proper calibration to the magnetic field readings from the Hall probe. Minimization of the sensor self-heating is ensured by passing a considerable low current of 0.25 mA. Interfacing to the platinum PT100 is done using a four-wire configuration rather than two-wire. This allows elimination of the effect of lead resistance as only the very low input bias current of the differential op amp passes through the two voltage readout terminals [21].

For proper temperature readouts, the nonzero offset voltage of the amplifier can be a problem as this drifts with time and temperature. In order to overcome this problem, the offset voltage is measured by reversing the current through the PT100 at a fixed frequency, in this case being 7.8 kHz.

When the current is reversed, the voltage due to the sensor reverses sign while thermal EMFs do not [22]. By averaging the forward and reverse current voltage measurements, the error in the voltage measurement due to thermal EMFs is thus eliminated.

3.2.5. Antialiasing Filter and Analogue to Digital Converter

The signal chain is kept fully differential from the Hall plate output to the ADC input. This provides increased immunity to external noise and better signal-to-noise ratio (SNR) performance. Also, a reduction in the even order harmonics is registered and a doubling in the dynamic range for the same voltage swing when compared to a single ended system is achieved. Figure 4 shows the differential signal levels at the Hall plate output and after amplification. For easier implementation the common mode voltage of the signal chain is kept tied to ground so as to avoid the introduction of additional offset voltages.

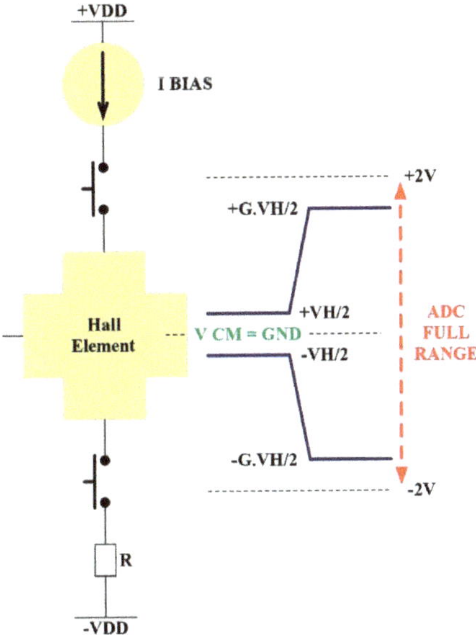

Figure 4. The differential signal levels at the Hall element output and after amplification.

A third-order low-pass Butterworth antialiasing filter with a designed bandwidth of 500 Hz, a quality factor of 0.707 and unity gain was designed and implemented for the necessary reduction of the out-of-band noise and minimization of distortion by matching the filter's output to the input circuitry of the ADC.

The multiple feedback circuit topology as shown in Figure 5 makes use of the two complex pole pairs in the feedback chain to set the desired cut-off frequency at 500 Hz whereas the third real pole at the output is set at a higher value in order to sum up an attenuation of −60 dB in the stop band [23].

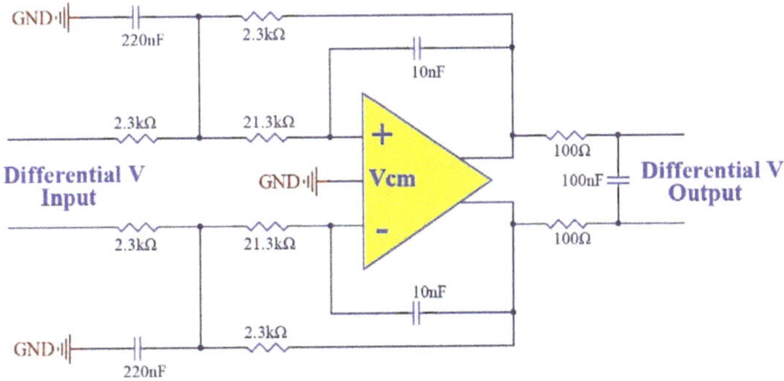

Figure 5. Third-order low-pass Butterworth differential antialiasing filter.

As the Butterworth filter topology offers the "maximally flat" response in the passband with the steepest roll-off, it was the preferred choice over other filter topologies.

Physical implementation of the Butterworth filters on the PCB involved optimal position placement of the passive components around the THS4131 differential op amp [24] in order to make the circuit as compact as possible and minimize the length of all trace runs for perfect symmetrical paths and minimal stray inductance pickup.

The antialiasing filter fully differential output is fed in the ADS131A04 24-bit delta-sigma analogue-to-digital converter [25]. The data rate flexibility, wide dynamic range and interface options makes this device well-suited for industrial and instrumentation applications where high precision digitization is required. A bipolar supply of ± 2.5 V is used to power up the analogue side and a separate power supply of +3.3 V derived from a separate low-dropout regulator (LDO) powers the digital side of the ADC. General EMC design guidelines were followed to keep the system noise as low as possible.

The external reference voltage (+4.096 V), which sets the input signal range, is derived from a separate reference source with heavy decoupling. This reference voltage is also internally buffered in the ADC for minimal loading effects.

The output data rate adjustment offers a tradeoff between the noise performance and the data acquisition frequency. When averaging is increased by reducing the data rate, noise drops considerably. As the system bandwidth is 500 Hz, the minimal data rate to be used according to Nyquist Sampling Theorem is 1 kHz.

The effective resolution defined for this ADC at the optimal noise performance data rate of 1 kHz is 22.19 bits. This is determined experimentally by shorting the analogue inputs together and taking an average of multiple readings across all channels. One second of consecutive readings are used to calculate the RMS noise [25]. The internal gain amplifier of the ADC is set to unity gain as this degrades the effective resolution. Equation (2) shows the relationship between the effective resolution and the RMS noise.

$$Effective\ Resolution = log_2\left(\frac{2 \times V_{REF}}{Gain \times V_{RMS}}\right) \tag{2}$$

Therefore 1.66 μV_{rms} results in 22.19 bits effective resolution. The analogue inputs of the ADC are directly connected to a switched-capacitor sampling network in an unbuffered mode. The ADC does not include any input buffers as these would induce input noise thus lowering the resolution. The capacitors of the switched capacitor input delta-sigma ADC are continuously being charged and discharged at the modulation sample frequency. Because the internal capacitors must be very small when compared to the external circuitry, the average input impedance of the ADC appears to be resistive. At a modulation frequency of 4.096 MHz and internal capacitor values of 3.5 pF, Equation (3) gives an input impedance of approximately 130 kΩ.

$$Z_{IN} = \frac{2}{f_{MOD} \times C_S} \tag{3}$$

The internal architecture of the delta-sigma ADC consists of a modulator at the input which samples the input signal at the rate of f_{MOD}. The modulator then converts the analogue input voltage into a pulse-code modulated (PCM) data stream. The $sinc^3$ digital filter takes this bitstream and provides attenuation to the now shaped higher frequency noise [25].

The magnitude frequency response of the $sinc^3$ filter has notches (or zeroes) that occur at the output data rate and its multiples. At these frequencies the filter has infinite attenuation. The $sinc^3$ filter magnitude frequency domain transfer function is given by Equation (4). As N is the decimation ratio which varies according to the set output data rate, the theoretical bandwidth of the filter depends on this frequency. Table 1 shows the filter's bandwidth for the set output data rates.

$$|H(f)| = \left|\frac{sin\left[\frac{N \times \pi \times f}{f_{MOD}}\right]}{sin\left[\frac{\pi \times f}{f_{MOD}}\right]}\right|^3 \tag{4}$$

Table 1. The sinc^3 filter bandwidth for set values of the output data rate.

Output Data Rate/kHz	Sinc^3 Filter Bandwidth/Hz
1	262
2	524
4	1048
8	2096

It can be seen that the desired 500 Hz bandwidth of the Hall probe is compromised at an acquisition frequency of 1 kHz. Predominantly due to the inclusion of the digital filter in the delta-sigma ADC chain, a calculable time delay between the analogue input and the digital output is present. This time delay is composed of the delay caused by the digital logic for the ADC to determine whether its conversions are synchronized. For this reason, the digital filter output is placed in a buffer for an entire conversion cycle before it is output. The second component of the delay is due to the group delay by the linear phase response of the sinc^3 filter as explained in [26].

The group delay of the sinc^3 filter is defined by Equation (5) where D is the decimation rate and f_M is the modulation frequency.

$$\tau_D = \left(\frac{D-1}{2}\right) \times \frac{3}{f_M} \tag{5}$$

Due to its linear phase response, the sinc^3 filter does not introduce additional distortion as no matter what the input frequency is, the output is always delayed by the same number of samples. Table 2 provides a breakdown of the group delay times for each output data rate.

Table 2. Breakdown of the time delays from the sinc^3 filter and digital logic of the ADC for set values of the output data rate.

Output Data Rate	1 kHz	2 kHz	4 kHz	8 kHz
OSR	4096	2048	1024	512
Number of Samples Delay from Sinc^3 filter/samples	1.4996	1.4993	1.4985	1.4971
Number of Samples Delay from Digital Logic/samples	1	1	1	1
Total Number of Samples Delay/samples	2.4996	2.4993	2.4985	2.4971
Total Group Delay/ms	2.4996	1.2496	0.6246	0.3121

Device communication is attained using a 20 MHz serial peripheral interface (SPI). The ADC is operated in continuous conversion synchronous master mode operation whereby the ADC signals the microcontroller of a complete data conversion by a negative edge trigger on the data ready (DRDY) line. The ADC then clocks out the last conversion data upon reception of the clock.

As data is transferred in 32-bit packet sizes and the analogue converted data is 24 bits long, the last 8 bits transferred of each packet are Hamming code validation bits. Calculation of the Hamming code on the received data is carried out by a software algorithm for each received packet and the data packet is scrapped in the scenario that the code received and calculated are not identical. Therefore, the Hamming code calculations on the digital interface to the ADC enhance the integrity of the communication channel.

3.2.6. Microcontroller and SDRAMs

The transfer of the acquisition data from the ADC is stored on board a 128 MB synchronous dynamic random access memory (SDRAM) during measurement time. The external memory interface module (EMIF) of the microcontroller supports a 32-bit data interface to four bank SDRAM devices. Communication is attained using the internal direct memory access module (DMA) of the microcontroller, which provides a hardware method of transferring data between peripherals and or/memory without intervention from the central processing unit (CPU), thereby freeing up the bandwidth of the CPU for other system functions. In this way, the ADC interrupt signal timing is not disrupted.

As 64 MB off-the-shelf SDRAM chips are available, two IS42S16320D-6BLI memory chips [27] are routed to the microcontroller whereby high-speed routing design techniques are applied for optimal performance.

The routing architecture of the two 54-ball TF-BGA SDRAM chips to the microcontroller assumes a symmetrical tree layout coupled with minimal clock skews between the command/address/control buses and the data bus [28].

An overall timing budget as explained in [29] was performed in order to determine the data-valid window considering a 100 MHz clock. The timing budget starts with the full cycle time allowed, in this case a 10 ns clock. As a general rule of thumb the total skew between the data lines should fall under 5% of the clock period, which gives an interval of 0.5 ns.

As the propagation delay of a microstrip line is given by Equation (6) in [30] and for a PCB with a typical dielectric constant ε_r of 4.4, the microstrip's delay constant results in 5.44 ps/mm.

$$t_{pd}(ps/mm) = \frac{85}{25.43} \times \sqrt{0.45\varepsilon_r + 0.67} \tag{6}$$

Therefore, the board skew for an interval of 0.5 ns is 91.69 mm. Matching the data trace lengths to a maximum length difference of 91.69 mm is not a problem in this case, as all trace lengths from the microcontroller to the SDRAM chips are all less than 60 mm in length. The transmitter and receiver skews are obtained from the device's data sheets and included in the timing budget.

As the high speed routing of the memory lines is mainly done from the two innermost signal layers which are sandwiched between two ground planes, micro strip transmission line theory is applied for the correct calculations of characteristic impedance and termination impedance matching [30].

For a signal trace of width W and thickness T, separated by distance H from a ground (or power) plane by a PCB dielectric with dielectric constant ε_r, the characteristic impedance is defined by Equation (7) with all measurement dimensions in mils.

$$Z_O(\Omega) = \frac{87}{\sqrt{\varepsilon_r + 1.41}} \times \ln\left[\frac{5.98 \times H}{(0.8 \times W + T)}\right] \tag{7}$$

Given that the trace width W is 0.1 mm, trace thickness T is 35 µm, dielectric thickness H of 135 µm and a relative dielectric constant ε_r of 4.4, the characteristic impedance of the traces is found to be 70.33 Ω.

Termination of the driver's output impedance to the transmission line is determined by finding the characteristic impedance of the source using the curves given in the input/output buffer specification (IBIS) model for the device driver which relates the inductance and capacitance of the pin and the silicon capacitance. The characteristic impedance varies slightly for each pin however all microcontroller pinouts connected to the SDRAMs were calculated to fall in the range of 35 to 45 Ω using Equation (8) as suggested in [31].

$$Z_T = \sqrt{\frac{L_{pin}}{C_{pin} + C_{comp}}} \tag{8}$$

Therefore, it was deemed best to terminate the driver's end of each transmission line so that the signals reflect off the unmatched end and terminate into the matched end as suggested in [32]. This was done by placing a series termination resistor of 22 Ω following the presented calculations at the driver's end of each trace. Data line signals being driven from both ends depending on a write or read command are terminated approximately in the middle of the line.

Other considerations were also taken in the routing and physical layout of the SDRAM chips. One-hundred nF ceramic decoupling capacitors are placed across the various power pins on the

SDRAM chips. This prevents the voltage supply from dropping when the SDRAM core requires current, as with a refresh, read or write. It also provides current during reads for the output drivers.

The number of vias on each line was also minimized in order to avoid adding extra capacitance on the traces. Also a keep out region around the microcontroller crystal was devised in order to prevent any high-speed routing across or close to the 16 MHz crystal.

3.2.7. Heidenhain Encoder Interface

As the end application of the instrument will be to map the magnetic field across the length of a 4-m-long undulator at PSI, the instrument supports an RS485-based interface to a Heidenhain linear absolute encoder with a resolution of 1 nm and an absolute accuracy of ± 3 μm. As the instrument is mounted on a rig mechanism along the undulator length, this industrial drive requires highly reliable and low-latency position feedback. The EnDat 2.2 protocol interface from HEIDENHAIN (Traunreut, Germany) is a digital bidirectional interface standard for position or rotary encoders. The interface transmits position values and also allows reading and writing of the encoder's internal memory. The type of data transmitted, like absolute position, temperature, diagnostic parameters and others, is selected through mode commands that the EnDat 2.2 master sends to the encoder [33].

Communication over the EnDat 2.2 interface with the encoder is implemented using a hardware configurable logic block module on the microcontroller that is accessed via library functions as explained in [34]. This block generates the clock for the encoder and for the internal SPI module that acts as the slave receiver and synchronizes communication with the encoder. Cable propagation delay compensation functions are also implemented via library functions.

Subsequent electronics circuitry consisting of RS-485 transceivers transmits differential data and clock signals in half-duplex mode and provide an end termination characteristic impedance of 120 Ω. The SN65HVD78 RS-485 transceivers [35] are chosen which can handle a maximum baud rate of 50 Mbps. This falls conveniently well above the maximum EnDat 2.2 protocol clock frequency of 16 MHz.

A very stringent requirement in this application is the synchronization of the magnetic field readings and the physical position reading across the undulator axis. The time duration between the falling edge of the ADC interrupt signal and the start of the encoder polling transmission command indicates the real time lag for the microcontroller to process the actual falling edge interrupt and enter in the programmed interrupt and set up the command in its registers to send to the encoder. This was determined experimentally as after the indicated lag of 8.4 μs in Figure 6 the data command is clocked out at which point the encoder saves its current position and later is clocked out and sent to the microcontroller. As the microcontroller operates at a clock frequency of 200 MHz, the microcontroller takes 1680 clock cycles to handle fully this request.

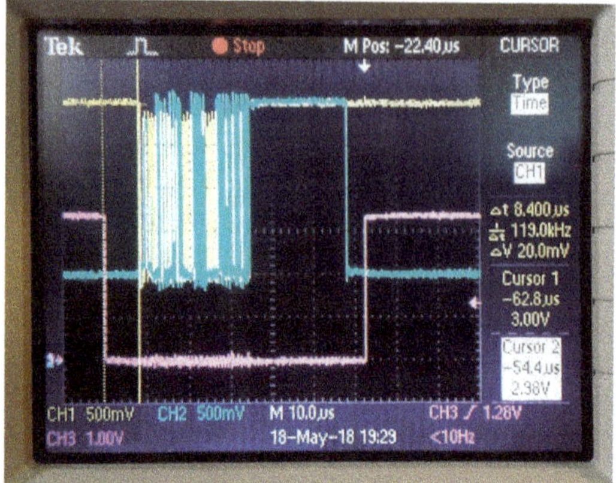

Figure 6. Oscilloscope snapshot showing the time delay of 8.4 µs between the falling edge of the ADC interrupt signal and the start of the encoder polling transmission command where the yellow trace is the encoder clock, blue trace is the encoder data and red trace is the ADC interrupt signal.

3.2.8. Voltage Regulation Circuitry

The instrument must be supplied with a ±6 V DC power supply. The 12 V across the supply rails directly feed a cooling fan placed on top of the microcontroller for convection heat extraction. The +6 V is regulated down to +5 V using the TPS7A4700 low dropout regulator [36]. This voltage rail powers the current source circuitry due to the headroom required by the op amp. The +5 V is then regulated to +3.3 V, which powers the positive rail of the remaining analogue circuitry. The +3.3 V is further regulated down to +2.5 V, which powers the positive supply rail of the analogue dynamic range of the ADC. This cascaded design of the LDOs (low dropout's) network was the preferred choice in improving drastically the power supply rejection ratio (PSRR) as enough voltage headroom is present between subsequent stages. Therefore as identical LDOs are used with a PSRR of 78 dB, the PSRR of the +2.5 V analogue supply voltage of the ADC triples to 234 dB, which provides excellent suppression of any noise and ripples from corrupting the ADC output.

The negative rail LDOs are also similarly cascaded providing comparable power supply noise performance for the analogue circuitry operating at −3.3 V and the negative ADC rail voltage of −2.5 V.

Because the total RMS supply current draw of the instrument is 300 mA and each LDO is capable of sourcing a 1 A load, negligible degradation in the PSRR is observed. The use of switching regulators for the generation of negative supply from positive supply is avoided in order to keep the system noise as low as possible.

The noise performance of the LDOs depends mostly on a noise free ground connection. Therefore, the thermal pad of each LDO is soldered directly to a pad on the PCB containing a 5 × 5 pattern of 0.25 mm vias for conducting heat. This thermal pad is also directly connected to the two internal ground planes.

3.2.9. Communication Interfaces: USB 2.0 and TTL Signals

The most convenient method of communication to the instrument is via a mini USB 2.0 port. The USB controller of the microcontroller operates as a full-speed function controller during point-to-point communications with the USB host. Due to the nature of data to be transferred, the bulk data transfer mode is implemented where communication is done using data bulk transfers of 512 bytes reaching a maximum transfer speed of 7 Mbit/s. Constrains in physical buffer sizes of the USB controller limit the maximum speed reached.

A bulk device class USB driver has been developed using the NI-VISA driver software by National Instruments Corporation (Austin, TX, USA) with the correct vendor ID and product ID of the instrument for device USB recognition. As the USB module is powered using a 60 MHz output clock set up by the auxiliary phase locked loop (PLL) and due to the differential signaling nature of the USB data lines, the D+ and D− traces from the microcontroller to the USB on board connector are precisely length matched to avoid any skews in data sent or received.

An external USB isolation dongle is provided with the instrument so that any ground return currents from the host side do not flow through the ground plane of the instrument. Such currents can induce ground voltage differentials that can affect the magnitude of the sensitive voltages of the analogue side. Also, isolation eliminates any voltage spikes on the ground that can occur on the host side during USB cable ejection and insertion.

Apart from control using the USB 2.0 interface, the instrument can be operated in standalone mode whereby control is exhibited only using a single transistor-transistor logic (TTL) external signal named "START/STOP".

Upon a low to high transition of this signal, the measurement process starts and a high to low transition stops the measurement process. Calibrated data is then stored on the SD Card and upon a high to low transition on the "BUSY" signal, data transfer is complete and the SD Card can be ejected. In the event of an error, the "ERROR" signal is pulled high.

These external signals are fed through an ISO7731 triple channel digital isolator [37] that provides an insulation barrier on both the supply and ground and the I/O pin of the microcontroller.

4. Embedded Microcontroller Software Program

The operation of the instrument is controlled by the 32-bit microcontroller. The program was developed and is stored on the flash read only memory (ROM). Upon power up, the CPU does its internal initialization routines and branches to the memory location of the starting point of the main code as defined in the linker file.

The flowchart shown in Figure 7 shows the sequential program flow that the instrument follows both in operation mode and calibration mode. The instrument defaults to operation mode after the microcontroller boots up and initializes all the on-board peripherals. Calibration mode must be chosen specifically by the user by sending a predefined instruction from the calibration LabVIEW (National Instruments Corporation, Austin, TX, USA) interface from the host PC through the USB connection. In this mode, real-time data is sent by the instrument through the USB so that the user performing calibration can graphically see real time plots and stores the data for analysis purposes.

The encoder is initialized only for operation mode and if no USB event connection is detected, the instrument enables the interrupts from the external TTL signals, otherwise these are disabled to avoid conflicting commands. In standalone mode the instrument temporary stores data on the SDRAM during measurement and after the "STOP" command is issued, the data is transferred automatically to the micro SD card. In USB connection mode, the instrument waits for a "START" command from the USB host and stores the acquired data on the SDRAM. After completion of the measurements, the user must select if data should be transferred to the USB host or on the micro SD card.

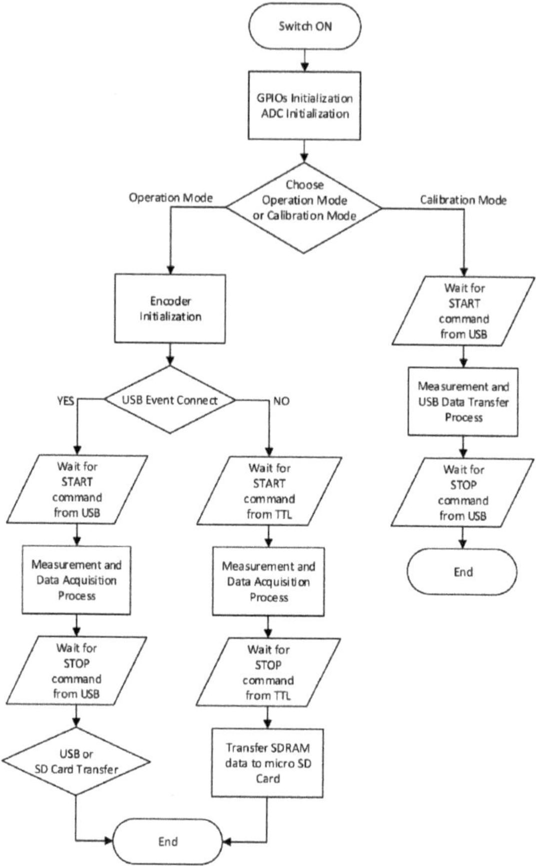

Figure 7. Flowchart of the instrument operation in both "Operation Mode" and "Calibration Mode".

5. Instrument Aluminum Enclosure

The instrument PCB is very sensitive and as both the top and bottom layers are populated with components, it must be handled with special care. A 1.6-mm-thick grey powder-coated aluminum enclosure was specifically designed and manufactured to house the PCB on 6-mm standoffs. The aluminum enclosure is also physically connected to the analogue ground plane and to the cable shielding of the Hall probe to provide electromagnetic shielding to the PCB internal to the enclosure.

A cooling fan is mounted on the inside of the enclosure to extract the heat mostly generated by the microcontroller and provide an airflow current through the enclosure for faster temperature stabilization of the electronics. Figure 8 shows the final instrument enclosure developed that houses the PCB. The air vent grille at the front serves as an air intake as the internal air circulation is extracted by the cooling fan.

Figure 8. The final developed instrument with external physical dimensions of the aluminum enclosure of 150 mm by 50 mm by 45 mm.

6. Experimental Measurements Results

6.1. Magnetic Field Range Testing

The instrument testing for the ± 2 T range is performed using the 7404 Lakeshore VSM electromagnet (Lake Shore Cryotronics, Westerville, OH, USA). This electromagnet is primarily used to characterize the DC magnetic properties of materials as a function of magnetic field, temperature and time. Through insertion of the Hall probe between the electromagnet poles, the sensor can be exposed to the full ± 2 T range and the instrument output response noted. The specifications of the 7404 Lakeshore VSM indicate a field accuracy of $\pm 0.05\%$ at full scale. The full scale at an air gap of 16.2 mm is specified to be ± 2.17 T resulting in an accuracy of ± 1mT.

Initial experimentation is performed in order to set the correct gain of the instrument analogue amplification stage so that full dynamic range of the ADC covers ± 2 T. Using standard value and high precision resistors with very low temperature coefficient, the gain of the amplification stage is set so that the ± 2 T reaches 93% of maximum signal swing to avoid saturation of the ADC analogue inputs. This is ensured by setting the magnetic field strength of the electromagnet to ± 2 T and tweaking the amplification voltage gain accordingly until a voltage of ± 3.809 V is read. This covers 93% of the full dynamic range of the ADC (± 4.096 V) and results in a final overall gain of 4.33 which gives a transduction ratio of 1.90464. Based on this ratio, noise performance results presented in the following sections are represented in Tesla rather than Volts.

6.2. Noise Performance

The noise performance of the whole instrument-sensor setup is determined using two figures of merits in order to quantify both the AC and the DC noise. Noise performance is characterized by the $1/f$ noise at quasi-DC measurement conditions and higher frequency noise beyond the $1/f$ corner frequency where the instrument is subject to white noise across the frequency spectrum up to the 500 Hz bandwidth of the Hall probe.

The DC resolution is given by the specification "Offset fluctuation and drift" whereas the AC resolution is given by the specification "Broadband" noise. The RMS noise voltage of the transducer in the frequency band from f_L to f_H is estimated by Equation (9) when combining both AC and DC noise.

$$V_{rmsB} \approx \left[NSD^2_{1/f} \times 1Hz \times ln\frac{f_H}{f_L} + 1.22 \times NSD^2_W \times f_H \right]^{\frac{1}{2}} \tag{9}$$

$NSD_{1/f}$ is the $1/f$ noise voltage spectral density at f equals to 1 Hz. NSD_W is the RMS white noise voltage spectral density. The numerical factor 1.22 is determined by a second-order low pass filter. To quantify both the AC and DC noise of the instrument for the different ADC output data rate frequencies, data acquisition of readily calibrated data is done over a time period of 10 s during which time the Hall Probe is placed in a high permeability three layer Mu-metal chamber that provides a near theoretical zero gauss test volume.

6.3. Offset Fluctuation and Drift (0.1–10 Hz)

The DC resolution of the instrument is limited by the offset fluctuation and drift in the frequency bandwidth from 0.1 to 10 Hz. Data acquired over a 10 s period is passed through an external digital second order low-pass Butterworth filter with a bandwidth of 10 Hz so that the out-of-band noise is largely attenuated. A 10 s period is taken in order to capture the full bandwidth from 0.1 to 10 Hz. Only the first term for $NSD_{1/f}$ is computed in Equation (9) in order to find the $1/f$ noise spectral density.

The standard deviation of the offset fluctuations corresponds to the integral noise of the device in the frequency range 0.1 to 10 Hz as explained in [38] and [39]. The optimal offset fluctuation and drift response is given at an output data rate of 1 kHz with a standard deviation noise figure of 0.78 µT. This degrades to 1.26 µT at 8 kHz output data rate. These results are shown in Table 3.

Table 3. Standard deviation and peak to peak noise performance figures for Offset Fluctuation and Drift.

Output Data Rate/kHz	Before 10 Hz External LPF		After 10 Hz External LPF		
	1σ Error/µT	Pk-Pk error/µT	1σ Error/µT	Pk-Pk error/µT	$NSD_{1/f}$/µT/$\sqrt{\mathrm{Hz}}$
1	1.770117	10.620700	0.782971	4.697829	0.364864
2	2.356910	14.141465	1.094589	6.567535	0.510077
4	2.635093	15.810560	1.160819	6.964918	0.540940
8	3.133331	18.799990	1.266418	7.598508	0.590149

6.4. Broadband Noise (10 Hz–f_T)

The AC resolution of the instrument is given by the specification "Broadband Noise". The calibrated acquisition data for the 10 s period is passed through a second order digital band-pass Butterworth filter with low and high cut off frequencies at 10 and 500 Hz respectively. This is done in order to capture the full bandwidth of the Hall probe, which is 500 Hz and filter out the "Offset fluctuation and drift" noise component. However, the internal sinc^3 filter of the ADC poses a frequency upper bandwidth limitation of 262 Hz when operated at 1 kHz output data rate. For the other data rates settings the bandwidth of the sinc^3 filter is beyond the 500 Hz bandwidth of interest.

The white noise spectral density is computed using the second term in Equation (9). An NSD_W of 0.063 µT/$\sqrt{\mathrm{Hz}}$ is achieved at an output data rate of 1 kHz which gives a standard deviation value of 1.56 µT. This defines the best AC noise performance of the instrument at a bandwidth of 262 Hz which degrades to 2.05 µT for the full sensor bandwidth of 500 Hz when operated at an output data rate of 2 kHz. Results are shown in Table 4 which are extracted from the histogram plots in Figure 9.

Table 4. Broadband noise performance figures for the indicated bandwidths.

Output Data Rate/kHz	Bandwidth/Hz	1σ Error/µT	NSD_W/µT/$\sqrt{\mathrm{Hz}}$
1	262	1.567720	0.0634
2	500	2.056552	0.0832
4	500	2.395707	0.0969
8	500	2.864964	0.1159

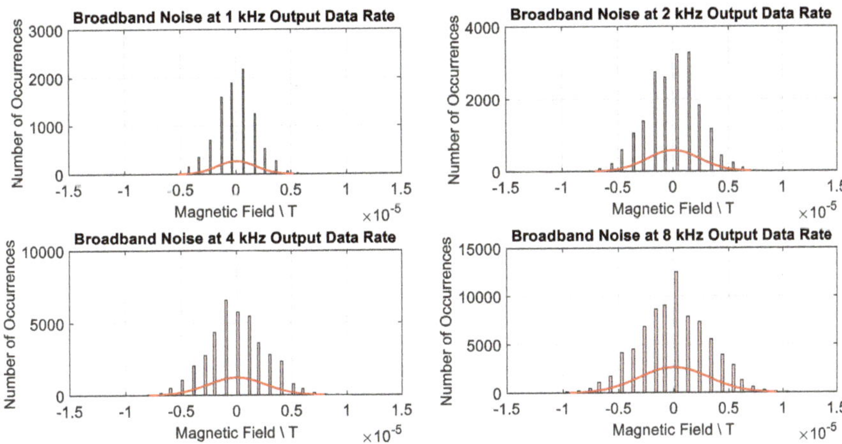

Figure 9. Broadband noise performance plots for the different output data rate settings. One can note that the DC component is filtered out by the external digital band pass filter.

7. Discussion

The theoretical magnetic resolution of the ADC of the newly developed instrument is 24 bits over the ±2 T calibration range. However, as the effective resolution of the ADC at the minimal output data rate of 1 kHz is 22.19 bits, this gives a magnetic resolution of 0.8 µT. The additional noise which results in a standard deviation of 1.2 µT is induced by the analogue interfacing circuitry, the Hall probe response and the induced noise in the Hall probe cable.

The SENIS H3A magnetic field transducer datasheet [9] has a specifications table that can be directly compared to the presented results in Section 6 for this instrument. The H3A transducer is specified to have a noise spectral density of 0.2 µT/$\sqrt{\text{Hz}}$ at $f = 1$ Hz and a noise spectral density of 0.05 µT/$\sqrt{\text{Hz}}$ at $f > 10$ Hz. These compare very well to the noise performance figures of the developed instrument which are 0.36 µT/$\sqrt{\text{Hz}}$ at $f = 1$ Hz and 0.063 µT/$\sqrt{\text{Hz}}$ at $f > 10$ Hz. It is to be noted and appreciated however, that since the SENIS H3A magnetic field transducer is fully analogue and external digitization to the instrument must be applied by an external analogue to digital data acquisition system, these noise figures only quantify the analogue noise read out as the noise spectral density measurements are done over a minimal range of ±100 mV with no amplification.

This is in contrast to the new developed instrumentation whose noise figures can be determined directly from the digital readings obtained from the on-board 24-bit ADC amplified to the ±2 T range as explained in Section 6. Hence whilst the former system only presents noise performance from the analogue stage, the latter system has comparable noise performance but includes both the analogue and the digital stage.

Bandwidth limitations are predominantly determined from the Hall probe sensor used, where highly nonhomogenous magnetic fields require integrated three-axis Hall probes whose frequency bandwidth ranges up to 75 kHz as pointed out in [39]. The SENIS Hall probe [7] used for mapping the Athos PSI line of undulators has a limited bandwidth of 500 Hz, which for this application is deemed to be sufficient. As the H3A magnetic field transducer [9] also interfaces to the same S type Hall probe, its bandwidth is also limited to 500 Hz.

Low-cost teslameters, such as the one presented in [40,41], provide a more cost-effective solution with limited range and accuracy. For example, a range of ±55 mT with an accuracy of 0.2% is covered by this teslameter. Low-cost commercially available teslameters such as this one are handheld instruments with performance relying on the limited 10-bit resolution of the built-in ADC of the MCU in this case. This makes such instrumentation unsuitable for very high precision field mapping.

Compared to these commercially available teslameters, the developed instrumentation provides a fully integrated solution with an optimized analogue-to-digital conversion stage and a proper spinning current analogue readout circuit for the interfacing of a three-axes Hall probe sensor. Hence, much higher performing circuitry is condensed in a smaller form factor volume.

Table 5 gives a summary of the main specifications of the best teslameters from three different leading vendors currently found on the market as indicated in [39]. The last column outlines the specifications of the developed teslameter presented here. It is to be noted that due to the on-chip integration of the signal conditioning electronics for the SENIS transducer, superior performance is achieved most notably in the bandwidth response. The published DC field accuracy defines the maximum difference between the actual measured magnetic flux density and that given by the teslameter. This is determined after a full calibration for the instrument is performed. The indicated DC field accuracy of 0.01% for the developed instrument is presented based on initial preliminary results after the application of nonlinearity calibration only which is modelled using a fifth order polynomial. This is deemed to be improved further by experimentation of higher order polynomials in this regard and the application of temperature calibration.

Table 5. General specifications summary of state of the art teslameters currently found on the market in comparison to the developed instrument. Information adapted from [39].

Parameter	SENIS	Projekt Elektronik	Group3	Developed Teslameter
DC Field Accuracy/%	0.002%	0.005%	0.01%	0.01%
Magnetic Resolution/μT	0.1 μT	0.1 μT	0.1 μT	0.8 μT
Measurement Range/T	1 μT–30 T	20 mT–2T	0.3–3T	$\pm$2 T
Bandwidth/kHz	DC–75 kHz	DC–1 kHz	DC–3 kHz	DC–0.5 kHz

An additional novelty of the developed electronic module is the quasi simultaneous measurement of both the magnetic field in the three axes and the absolute linear encoder position with only an 8.4 μs delay lag. Even though the instrument has been developed as a general instrument that can be used for other machines, its specification has been developed, first and foremost, for the ATHOS soft X-Ray beamline at PSI in order to achieve a true magnetic field map across the length of its undulators. When considering that the developed instrument has similar performance to the state-of-the-art teslameters [39] but with a much smaller form factor, an integrated digitization stage, integrated position reading capability and ultimately a cheaper solution—one can appreciate that this new instrument offers several advantages over the state of the art for use in other machine undulator mapping applications and even general purpose ones.

Additional research is currently being conducted in order to devise the best calibration methods for nonlinearity calibration, temperature changes compensation and Hall probe angular errors correction.

Temperature compensation will surely include offset and sensitivity compensation over the full $\pm$2 T dynamic range for the precise modelling of the effect of the temperature changes on the magnetic field readout. As indicated in [2], the offset and sensitivity of the Hall device drift over temperature and this varies randomly from one device to another. These drifts are mostly caused by piezo-resistive effects. Also, the mechanical stress induced by thermal expansion of the material results in such effects.

The influence of temperature compensation of sensitivity will have a much greater effect than the compensation for the temperature offset only at zero Gauss. Temperature compensation of sensitivity will model the temperature effect at different magnetic field values across the whole range. Modelling this effect will include second or higher order polynomials in order not to degrade the accuracy obtained after non linearity calibration. Also, as this Hall probe is specified to have an orthogonality error of <2°, this will also be calibrated.

8. Conclusions

The developed high-accuracy teslameter can measure magnetic fields in the range of $\pm$2 T. The instrument was designed within the required specifications set by the PSI undulator characterization and provides digitized readings of the three magnetic field axes, the PT100 and the absolute encoder position readings—all synchronized together.

Rigorous testing is performed using output data rates ranging from 1 to 8 kHz with the best noise performance results achieved at 1 kHz with a standard deviation of 1.56 µT at a limited bandwidth of 262 Hz. Measures for resolution are given by the offset fluctuation and drift with a noise spectral density of 0.36 µT/$\sqrt{\text{Hz}}$ and a broad band noise spectral density of 0.06 µT/$\sqrt{\text{Hz}}$.

Work is underway to establish an optimized and standardized calibration setup for the instrument in order to maximize its performance and measurement accuracy in micro Tesla.

Author Contributions: Formal analysis, J.C.; Investigation, J.C.; Methodology, J.C.; Software, J.C.; Supervision, A.S. and N.S.; Validation, A.S., N.S. and M.C.; Writing—original draft, J.C.; Writing—review & editing, A.S., N.S., M.C., S.D. and R.S.P.

Funding: This research received no external funding.

Acknowledgments: The authors would like to thank Reuben Debono for his useful guidance and help in the PCB assembly of the instruments at the Electronic Systems Lab at the Faculty of Engineering at University of Malta. The authors would like to thank R. Ganter, project leader of the Athos undulator beamline and H-H. Braun, SwissFEL machine director, for their constant support throughout the entire project. The authors would like to thank Sasa Spasic and his team at Sentronis facilities for their fruitful discussions and their guidance during testing.

Conflicts of Interest: The authors declare no conflicts of interest.

References

1. Satav, S.M.; Agarwal, V. Design and Development of a Low-Cost Digital Magnetic Field Meter with Wide Dynamic Range for EMC Precompliance Measurements and other Applications. *IEEE Trans. Instrum. Meas.* **2009**, *58*, 2837–2846. [CrossRef]

2. Sanfilippo, S. Hall probes: Physics and application to magnetometry. In Proceedings of the Specialised Course on Magnets, Bruges, Belgium, 16–25 June 2009; pp. 423–462.

3. Popovic, D.R.; Dimitrijevic, S.; Blagojevic, M.; Kejik, P.; Schurig, E.; Popovic, R.S. Three-Axis Teslameter With Integrated Hall Probe. *IEEE Trans. Instrum. Meas.* **2007**, *56*, 1396–1402. [CrossRef]

4. Tosin, G.; Citadini, J.; Conforti, E. Hall-Probe Bench for Insertion-Device Characterization at LNLS. *IEEE Trans. Instrum. Meas.* **2007**, *56*, 2725–2730. [CrossRef]

5. Gehlot, M.; Mishra, G. Development of Stretched wire measurement bench at IDDL, DAVV Indore. *J. Phys. Conf. Ser.* **2016**, *755*, 012054. [CrossRef]

6. Calvi, M.; Brugger, M.; Danner, S.; Imhof, A.; Johri, H.; Schmidt, T.; Scoular, C. Swissfel U15 Magnet Assembly: First Experimental Results. In Proceedings of the FEL2012, Nara, Japan, 26–31 August 2012; pp. 662–665.

7. Hall Probe S for H3A Magnetic Field Transducers. Available online: http://c1940652.r52.cf0.rackcdn.com/57971209b8d39a2071000bc2/hall-probe-s-for-h3a-mft-datasheet-r2[1].pdf (accessed on 26 March 2019).

8. Paul Scherrer Institut. SwissFEL Concept Design Report. Available online: https://inis.iaea.org/collection/NCLCollectionStore/_Public/42/006/42006326.pdf?r=1&r=1 (accessed on 26 March 2019).

9. H3A Magnetic Field Transducer. Available online: http://c1940652.r52.cf0.rackcdn.com/5791ded9b8d39a2071000a9c/h3a-magnetic-transducers-datasheet_rev02.pdf (accessed on 26 March 2019).

10. Popovic, R.S. High Resolution Hall Magnetic Sensors. In Proceedings of the 29th International Conference on Microelectronics, Belgrade, Serbia, 12–14 May 2014; pp. 69–74.

11. Mosser, V.; Matringe, N.; Haddab, Y. A Spinning Current Circuit for Hall Measurements down to the Nanotesla Range. *IEEE Trans. Instrum. Meas.* **2017**, *66*, 637–650. [CrossRef]

12. Junfeng, J.; Wilko, K.; Kofi, M. A Continuous-Time Ripple Reduction Technique for Spinning-Current Hall Sensors. *IEEE J. Solid-State Circuits* **2014**, *49*, 1525–1534.

13. Data Sheet for TMS320F2837xD Dual-Core Delfino™ Microcontrollers. Available online: http://www.ti.com/lit/ds/symlink/tms320f28379d.pdf (accessed on 26 March 2019).

14. PCB Design Guidelines for reduced EMI. Available online: http://www.ti.com/lit/an/szza009/szza009.pdf (accessed on 26 March 2019).

15. High-Speed Interface Layout Guidelines. Available online: http://www.ti.com/lit/an/spraar7h/spraar7h.pdf (accessed on 26 March 2019).

16. Design Considerations for Mixed-Signal PCB Layout. Available online: https://www.teledyne-e2v.com/content/uploads/2014/09/Board-Layout.pdf (accessed on 26 March 2019).
17. Data Sheet for ADG1611/ADG1612/ADG1613. Available online: https://www.analog.com/media/en/technical-documentation/data-sheets/adg1611_1612_1613.pdf (accessed on 26 March 2019).
18. Data Sheet for INA2128 Dual, Low Power Instrumentation Amplifier. Available online: http://www.ti.com/lit/ds/symlink/ina2128.pdf (accessed on 26 March 2019).
19. AN-1515 A Comprehensive Study of the Howland Current Pump. Available online: http://www.ti.com/lit/an/snoa474a/snoa474a.pdf (accessed on 26 March 2019).
20. Data Sheet for REF50xx Low-Noise, Very Low Drift, Precision Voltage Reference. Available online: http://www.ti.com/lit/ds/symlink/ref5050.pdf (accessed on 26 March 2019).
21. AN-1559 Practical RTD Interface Solutions. Available online: http://www.ti.com/lit/an/snoa481b/snoa481b.pdf (accessed on 26 March 2019).
22. Appendix E: Temperature Measurement System. Available online: https://www.lakeshore.com/Documents/LSTC_appendixE_l.pdf (accessed on 26 March 2019).
23. Fully-Differential Amplifiers. Available online: http://www.ti.com/lit/an/sloa054e/sloa054e.pdf (accessed on 26 March 2019).
24. Data Sheet for THS413x High-Speed, Low-Noise, Fully-Differential I/O Amplifiers. Available online: http://www.ti.com/lit/ds/slos318i/slos318i.pdf (accessed on 26 March 2019).
25. Data Sheet for ADS131A0x 2- or 4-Channel, 24-Bit, 128-kSPS, Simultaneous-Sampling, Delta-Sigma ADC. Available online: http://www.ti.com/lit/ds/symlink/ads131a04.pdf (accessed on 26 March 2019).
26. Accounting for delay from multiple sources in delta-sigma ADCs. Available online: http://www.ti.com/lit/wp/slyy095a/slyy095a.pdf (accessed on 26 March 2019).
27. Data Sheet for IS42/45R86400D/16320D/32160D IS42/45S86400D/16320D/32160D. Available online: http://www.issi.com/WW/pdf/42-45R-S_86400D-16320D-32160D.pdf (accessed on 26 March 2019).
28. DDR2 (Point-to-Point) Package Sizes and Layout Basics. Available online: https://www.micron.com/support/~{}/media/c40b5a5e4afd4bdfa285edf8e5ca1017.ashx (accessed on 26 March 2019).
29. DDR SDRAM Point-to-Point Simulation Process. Available online: https://www.micron.com/~{}/media/documents/products/technical-note/.../tn4611.pdf (accessed on 26 March 2019).
30. Microstrip and Stripline Design. Available online: https://www.analog.com/media/en/training-seminars/tutorials/MT-094.pdf (accessed on 26 March 2019).
31. The IBIS model, Part 3: Using IBIS models to investigate signal-integrity issues. Available online: http://www.ti.com/lit/an/slyt413/slyt413.pdf (accessed on 26 March 2019).
32. Termination for Point-to-Point Systems Introduction. Available online: https://www.micron.com/-/media/client/.../tn4606_point_to_point-termination.pdf (accessed on 26 March 2019).
33. EnDat 2.2 – Bidirectional Interface for Position Encoders. Available online: http://www2.asiaa.sinica.edu.tw/~{}homin/wiki/pmwiki-2.1.27/uploads/ALMA/Endat2_1and2_2.pdf (accessed on 26 March 2019).
34. C2000 Position Manager EnDat22 Library Module User Guide. Available online: http://www.ti.com/lit/ug/sprui35/sprui35.pdf (accessed on 26 March 2019).
35. Data Sheet for SN65HVD7x 3.3-V Supply RS-485 With IEC ESD Protection. Available online: http://www.ti.com/lit/ds/sllse11h/sllse11h.pdf (accessed on 26 March 2019).
36. Data Sheet for TPS7A470x 36-V, 1-A, 4-µVRMS, RF LDO Voltage Regulator. Available online: http://www.ti.com/lit/ds/symlink/tps7a47.pdf (accessed on 26 March 2019).
37. Data Sheet for ISO773x High-Speed, Robust-EMC Reinforced Triple-Channel Digital Isolators. Available online: http://www.ti.com/lit/ds/symlink/iso7731.pdf (accessed on 26 March 2019).
38. Popovic, D.R.; Dimitrijevic, S.; Spasic, S.; Popovic, R.S. High-accuracy teslameter with thin high-resolution three-axis Hall probe. *Measurement (Lond.)* **2015**, *98*, 407–413. [CrossRef]
39. Renella, D.P.; Spasic, S.; Dimitrijevic, S.; Blagojevic, M.; Popovic, R.S. An overview of commercially available teslameters for applications in modern science and industry. *Acta IMEKO* **2017**, *6*, 43–49. [CrossRef]

40. Jovanovic, U.; Jovanovic, I.; Blagojevic, M.; Krstic, D.; Mancic, D. Low-cost Teslameter based on Hall Effect Sensor MLX90242. *Serb. J. Electr. Eng.* **2018**, *15*, 225–232. [CrossRef]
41. Jovanovic, U.; Jovanovic, I.; Blagojevic, M.; Mancic, D. One Solution of a Low-Cost Teslameter. In Proceedings of the 13th International Conference on Applied Electromagnetics, Nis, Serbia, 30 August–1 September 2017.

MDPI

St. Alban-Anlage 66

4052 Basel

Switzerland

Tel. +41 61 683 77 34

Fax +41 61 302 89 18

www.mdpi.com

Electronics Editorial Office

E-mail: electronics@mdpi.com

www.mdpi.com/journal/electronics

www.ingramcontent.com/pod-product-compliance
Lightning Source LLC
LaVergne TN
LVHW070920170726
843515LV00005B/828